Advances in Anatomy, Embryology and Cell Biology

Volume 241

Advances in Anatomy, Embryology and Cell Biology publishes critical reviews and state-of-the-art research in the areas of anatomy, developmental and cellular biology. Founded in 1891, this book series has a long standing tradition of publishing focused and condensed information on a given topic with a special emphasis on biomedical and translational aspects.

The series is open to both contributed volumes (each collecting 7 to 15 focused reviews written by leading experts) and single-authored or multi-authored monographs (providing a comprehensive overview of their topic of research).

Advances in Anatomy, Embryology and Cell Biology is indexed in Medline, SCImago, SCOPUS.

Pei-Shiue Tsai
Editor

Extracellular Vesicles in Reproduction

Origin, Functions and Perspective

Springer

Editor
Pei-Shiue Tsai
School of Veterinary Medicine
National Taiwan University
Taipei, Taiwan

ISSN 0301-5556 ISSN 2192-7065 (electronic)
Advances in Anatomy, Embryology and Cell Biology
ISBN 978-3-032-04488-4 ISBN 978-3-032-04489-1 (eBook)
https://doi.org/10.1007/978-3-032-04489-1

This Springer imprint is published by the registered company Springer Nature Switzerland AG
The registered company address is: Gewerbestrasse 11, 6330 Cham, Switzerland

Preface

The Importance of Exosomes in Reproductive Biology

Exosomes, small extracellular vesicles ranging from 30 to 150 nm, have emerged as pivotal mediators of intercellular communication across various physiological systems, including the reproductive system. Their significance in reproductive biology has garnered increasing attention, particularly due to their roles in modulating sperm maturation, sperm/oocyte transportation, embryo development, and maternal-fetal interactions. Derived from almost all cell types, exosomes carry a rich cargo of proteins, lipids, mRNAs, and microRNAs (miRNAs), which are transferred to recipient cells to modulate gene expression and cellular function.

In male reproduction, exosomes are secreted by the epididymal epithelium and play a critical role in sperm maturation. These epididymal exosomes, also known as epididymosomes, deliver essential proteins and regulatory RNAs to developing spermatozoa, contributing to their motility and fertilization capacity. Furthermore, in seminal plasma, exosomes influence immune tolerance in the female reproductive tract post-coitus, reducing the likelihood of immune rejection of the male gametes. In female reproduction, follicular fluid-derived exosomes have been shown to regulate oocyte maturation and cumulus cell function. These vesicles can carry miRNAs that influence ovarian follicle development and granulosa cell proliferation. Similarly, exosomes present in the uterine fluid support embryo implantation and early embryonic development by modulating endometrial receptivity and cell signaling pathways. Additionally, during early pregnancy, trophoblast-derived exosomes interact with maternal immune and endometrial cells to facilitate implantation and modulate immune tolerance. These vesicles may help orchestrate a local immune environment that allows the semi-allogenic embryo to be accepted by the maternal immune system. Moreover, placental exosomes carry signals that regulate angiogenesis, inflammation, and even fetal growth, indicating their far-reaching systemic effects. All these above-mentioned points clearly demonstrate that exosomes are not merely passive cellular by-products but active participants in reproductive processes. They contribute to the fine-tuned molecular dialogue required for successful fertilization, implantation, and pregnancy maintenance. As our understanding of exosomal biology advances, new frontiers may open in fertility preservation, embryo selection, and the treatment of reproductive pathologies.

Therefore, in this volume, we cover recent findings regarding exosomes detected within the reproductive system; introduce their essential roles in the regulation of sperm maturation (epididymosomes), sperm activity (seminal vesicle-derived exosomes), and embryo-endometrium communication (uterosomes); and aim to contribute to the continued research into the roles and mechanisms of exosomes in reproductive biology not only scientifically compelling but also of profound clinical relevance.

Taipei, Taiwan Pei-Shiue Tsai

Contents

Extracellular Vesicles, Nutrient Balance, and Redox Metabolism During Epididymal Sperm Maturation: *A Perspective on the Complexity of Intercellular Interplay*

Bao Li Zhang, Bao Ying Liu, Shang Cao, Shuo Shi, and Winnie Shum

Contents

B. L. Zhang
USJ-Kong Hon Academy for Cellular Nutrition and Health, Faculty of Health Sciences, University of Saint Joseph, Macao, China

NHC Key Lab of Reproduction Regulation, Shanghai Engineering Research Center of Reproductive Health Drug and Devices, Shanghai Institute for Biomedical and Pharmaceutical Technologies, Shanghai, China

B. Y. Liu and S. Cao
School of Life Science and Technology, ShanghaiTech University, Shanghai, China

S. Shi
Shanghai Institute for Advanced Immunochemical Studies, ShanghaiTech University, Shanghai, China

W. Shum (✉)
USJ-Kong Hon Academy for Cellular Nutrition and Health, Faculty of Health Sciences, University of Saint Joseph, Macao, China

NHC Key Lab of Reproduction Regulation, Shanghai Engineering Research Center of Reproductive Health Drug and Devices, Shanghai Institute for Biomedical and Pharmaceutical Technologies, Shanghai, China

School of Life Science and Technology, ShanghaiTech University, Shanghai, China

Institute of Science and Environment, University of Saint Joseph, Macao, China
e-mail: winnie.shum@usj.edu.mo; shumw@shanghaitech.edu.cn

Adv Anat Embryol Cell Biol (2025) 241: 1–87
https://doi.org/10.1007/102_2025_10,

Published Online: 09 October 2025

Abstract Life relies on redox reactions to sustain; however, imbalanced redox exerts stress on living beings. This is particularly pivotal as mammalian gamete sperm cells undergo functional maturation in the epididymis, preparing themselves for the long and challenging journey to begin a new life by successfully fertilizing an oocyte. Defects in epididymal sperm maturation are closely related to male infertility and reproductive health. In the epididymis, the halted translational machinery in spermatozoa while maturing in a quiescent state prompts them to immensely depend on epithelial cells for nutritional supports and information exchange. Extensive intercellular communication is therefore required between sperm and epithelial cells, and extracellular vesicles (EVs) play a crucial role as intercellular carriers. Epididymal luminal microenvironment is spatially specialized to be an acidic, decalcitonic, and pro-reductive protective milieu to prevent premature sperm dysfunction. The uniqueness of epididymal microenvironment also requires the EVs to have adaptive characteristics. Although the biogenesis and compositions of EVs in the epididymis have been reviewed elsewhere, this chapter discusses the potential role of EVs from the perspective of cellular nutrient balance of vitamins and minerals, redox metabolism, and intercellular interplay during sperm maturation to better understand the biology of epididymis in male reproduction.

Keywords Acid-base balance, Epididymal sperm maturation, Epithelial cells, Extracellular vesicles, Intercellular carriers, Male reproduction, Matrix calcium, Micronutrients, Mineral metabolism, Redox metabolism

Key Points

- A literature update on the function of epithelial cells in the epididymis in conferring sperm maturation, focusing on the pivotal role of extracellular vesicles (EVs) as intercellular carriers.

- A brief literature analysis on the redox metabolism in epididymal function and the characteristics of the adaptation of EVs to the pro-reductive acidic and decalcitonic microenvironment, and their contributions to the redox metabolic regulation during sperm maturation.
- A brief literature summary of the redox properties of nutrients in cellular metabolism, including vitamins and essential elemental minerals, and their participation in regulation of epididymal function and reproductive health based on the existing literature.

1 General Introduction

It is a long and challenging journey for mammalian haploid sperm to encounter and successfully fertilize another haploid oocyte and begin a new life. The production of viable and intact sperm is a fundamental prerequisite for this reproductive process to occur. Male reproduction is achieved by normal spermatogenesis in the testis and functional maturation of sperm in the epididymis (Gervasi and Visconti 2017; Orgebin-Crist 1967, 1969). Epididymis is one highly convoluted single-tubular organ lined with a layer of specialized types of epithelial cells and connects the testis via efferent ducts to the vas deferens. Intercellular communication between somatic epididymal cells in the reproductive tract and gametic sperm cells is crucial, and intercellular carriers, extracellular vesicles (EVs), are particularly important. Epididymal-dependent sperm maturation plays an indispensable role in subsequent functional activities of sperm, including swimming into the female tract, undergoing capacitation, hyperactivation and acrosome reactions, sperm-egg interactions, and eventually, the initiation of a healthy embryo that develops the life of next generation. These post-testicular reproductive events highlight the essential role of epididymal function and thereby the role of EVs in sperm maturation. While our understanding of post-testicular reproductive biology remains to be elucidated, epididymal dysfunction has received increasing attention due to its association with offspring health (Chen et al. 2016a, b; Sharma et al. 2016) and male infertility, especially idiopathic infertility and reproductive defects (Agarwal et al. 2021; Punab et al. 2017; Wang et al. 2024b; WHO 2017). However, there is still no cure for male infertility, which is affecting as many as or even more than 15% of couples of childbearing age (Belladelli et al. 2023). Infertile couples seek help from assisted reproductive technology, whereas exploring alternative approaches to address the problem from the basis is still needed. A better understanding of the biology of epididymis may help us address this public health problem of infertility.

Epididymal function is conferred by a layer of epithelial cells lined at the front interface of luminal cavity, which is filled with epididymal fluid, in which sperm mature and are nourished at a quiescent state while protected from immune challenges. Testicular sperm are functionally immature and obtain their mature fertilization competence while transiting in the epididymis, an essential process to ensure male fertility and fecundity. Anatomically, the epididymis is made up of three distinct regions: the caput (head), which receives the spermatozoa from the efferent ducts; the corpus (body); and the cauda (tail), where spermatozoa are stored and pass to the vas deferens. Depending on the species, for example, in seasonal mammals, spermatozoa require at least about 2 days to pass through the epididymis and can be stored in the cauda epididymidis up to several months (Amann and Howards 1980; Hermo and Robaire 2002; Robaire and Hinton 2015). The epididymal tubular mucosa comprises different types of cells, including epithelial principal cells, clear/narrow cells and basal cells, immunological cells from immune system, and smooth muscle cells that surround the tubule outside the epithelial layer (Breton et al. 2019; Elfgen et al. 2018; Hermo et al. 2025; Robaire and Hinton 2015; Shum et al. 2022). These various types of cells are specially organized and worked in a concerted manner to respond to physiological challenges and maintain the epididymal function in a normal state. The epididymal epithelial cells establish a specialized and protective luminal fluid in a regional-dependent manner, for example, an acidic and decreasing calcium gradient milieu, so that spermatozoa mature in a physiologically and immunologically "unique" microenvironment down the epididymal tubule. In the epididymal microenvironment, sperm undergo an immense, but tightly controlled, spatiotemporal maturation process, particularly the process of remodeling on the cell surface (Belleannee et al. 2011; Breton et al. 2019; Cheung et al. 2005; Gervasi and Visconti 2017; Rejraji et al. 2006; Robaire and Hinton 2015; Shum et al. 2022; Sullivan et al. 2005; Wong et al. 2002; Wu et al. 2017; Zhou et al. 2018). The transcription machinery of sperm cells in the epididymis is paused due to cell cycle arrest and highly condensed chromatin (Braun 1998; Ward 2009), so they require nutritional support and protection from the epithelium. The epididymal epithelium at the front line in contact with spermatozoa, directly or indirectly, is therefore responsible for the extensive exchange activities between the gametic sperm cells and the somatic epithelial cells, as well as the formation of luminal fluid. The epididymal luminal fluid is a mixture produced by epithelial cells with contributions from upstream testicular fluid and intrinsic factors in the spermatozoa, and the fluid in turn serves as the media supporting extensive intercellular exchange activities. The EVs, also known as epididymosomes in the epididymis, are potent messengers of intercellular communication due to their ability to transfer genetic matter DNAs and RNAs, enzymes, and nutrients, including proteins and lipids (Barrachina et al. 2022; Belleannee et al. 2013; Chan et al. 2020; Frenette et al. 2003, 2010; James et al. 2020; Rejraji et al. 2006; Sullivan et al. 2005). They contribute profoundly the constituents of epididymal fluid and mutually influence the dynamic processes of gametic-somatic intercellular communication, subsequently playing a crucial role in epididymal function and sperm maturation.

The production and release of EVs are evolutionarily conserved processes that play critical roles in various vesicle-mediated physiological functions, particularly in the intercellular transport of RNA (Jeppesen et al. 2019, 2023; Skog et al. 2008; Tkach and Thery 2016; Valadi et al. 2007; Wolf 1967). This underscores the importance of EVs biology in epididymal function and sperm maturation, building on the pioneering work by Sullivan and coworkers on the exchange of proteins in sperm with EVs (Frenette et al. 2002; Frenette and Sullivan 2001) and subsequent studies on the exchange of lipids and RNAs (Ali et al. 2023; Barrachina et al. 2022;

Belleannee et al. 2013; Chan et al. 2020; Conine et al. 2018; Fazzio et al. 2024; Frenette et al. 2003, 2010; Moon et al. 2024; Rejraji et al. 2006; Sharma et al. 2016, 2018; Sullivan et al. 2005). Accordingly, EVs are membranous, lipid-bilayer enclosed vesicles released by cells (Jeppesen et al. 2023). They are generically classified into subsets based on their size, including small EVs (<200 nm), small to large EVs (100 nm—5,00 nm), and large EVs (>500 nm–10 μm). With consideration of biogenesis and specific markers, EVs can be further subdivided into several subclasses: exosomes are well-known small EVs with sizes ranging from 40 to 100 nm, whereas microvesicles (MVs), apoptotic bodies, and microparticles usually belong to the category of EVs with larger sizes. Beyond membrane-bound vesicles, membrane-free nanoparticles (NPs) constitute another major class of extracellular carriers (Jeppesen et al. 2023). However, the landscape of these two superfamilies of extracellular carriers remains incomplete. This is due to the diversity and heterogenicity of these carriers with a substantial overlap in their physicochemical properties, as well as the limitations of the investigation methods that are mainly based on the sample preparation according to size and gravity (Das et al. 2019; Makarova et al. 2023). The complexity of the biology of extracellular carriers in transporting extracellular RNAs is also due in part to the fact that early findings on these extracellular carriers may have associated with multiple subtypes of these intercellular carriers. Recent studies have shown that proteins and nucleic acids previously thought to be packaged in exosomes appear to be more enriched in different types of EVs and in membraneless NPs (Jeppesen et al. 2019; Jeppesen et al. 2023; Lasser 2019; Murillo et al. 2019). Some NPs, also called stress granules, are also known to play an important role in sperm maturation in the epididymis (Ma et al. 2019) and in male fertility (Wang et al. 2024a). In view of the fact that the biological field of membrane-free NPs is relatively new and the challenges of unambiguous characterization remain to be explicated (Jeppesen et al. 2023; Makarova et al. 2023; Yanez-Mo et al. 2015), EVs have relatively longer history since their first discovery by electron microscopy in 1967 as a product of platelets (Wolf 1967), followed by the discovery that EVs can transport RNA between cells (Skog et al. 2008; Valadi et al. 2007). Therefore, in this chapter, we focus only on the biological role of membrane-enveloped EVs as extracellular carriers during epididymal sperm maturation.

The microenvironment within the epididymal lumen provides a balanced nutrient fluid but is continuously challenged by the stress generated by the spermatozoa undergoing maturation in the epididymis. In the testes of adult men, millions of spermatozoa are released into the testicular fluid each day and reach the initial segment of the epididymis (Amann and Howards 1980; Skakkebaek et al. 2022). During the transit of spermatozoa through the epididymis, they have two basic needs, metabolism and remodification, the latter referring to the changes in their membranes and coats and post-translation modifications. However, in spermatozoa, histones are replaced by protamines to compress DNA to higher order, making the classical processes of transcription and translation impossible. And, therefore, spermatozoa must rely on the microenvironment that supports them for the nurtures and exchange of their metabolic wastes. The epididymal luminal fluid is specifically constituted in a spatial-dynamical manner that is different from blood or body fluids to fulfil the dynamic homeostatic demands (i.e., homeodynamics (Lloyd et al. 2001)) of sperm transiting the epididymis for maturation. For example, the luminal fluid is mildly acidic, has a gradual decrease in calcium gradient, i.e., decalcitonic, and is highly viscous and specifically enriched with essential organic and inorganic nutrients (Breton and Brown 2013; Hermo and Robaire 2002; Levine and Marsh 1971; Myers and Cornwall 2024; Shum et al. 2022; Turner 2002). Distinct kinds of RNAs have also been reported in the epididymis, which play an important role in male fertility and transgenerational inheritances (Chen et al. 2016b; Conine et al. 2018; Dhakal et al. 2019; Sharma et al. 2016, 2018). In addition, a small dose of redox inputs is also required to promote sperm maturation (Aitken and Vernet 1998; Baskaran et al. 2020a; O'Flaherty 2019); however, oxidative stress is detrimental to epididymal function and sperm health (Dutta et al. 2019). This chapter therefore aimed to focus on the function of EVs from the perspective of nutritional and redox balance, and the intercellular interplay based on these two aspects during epididymal sperm maturation. While a comprehensive discussion is not intended in this chapter, we give our opinion and perspectives on these two aspects of EVs in the epididymal function and provide some explorative thoughts that may be useful for future research in this field. For more background details about epididymal function, the readers can refer to the textbook by Hermo and Robaire (2002) and to other informative reviews, including the biochemical and molecular physiological aspects of epididymis (Dacheux and Dacheux 2014; Robaire and Hinton 2015), acidification and intercellular interactions (Breton and Brown 2013; Cheung et al. 2005; Leung et al. 2004; Pastor-Soler et al. 2005; Shum et al. 2009), epididymal matrix amyloid in host defense (Cornwall et al. 2019; Myers and Cornwall 2024), epididymal cargo and epididymosomes (Ali et al. 2023; Fazzio et al. 2024; James et al. 2020; Mear et al. 2024; Sullivan et al. 2005), epididymal immunity (Battistone et al. 2024; Zhao et al. 2020), epithelial cellular function (Breton et al. 2019; Chen et al. 2021), luminal nutrients and microenvironment (Shum et al. 2022; Zhou et al. 2018), molecular changes of epididymal sperm (Gervasi and Visconti 2017), pathological aspects (Weigel Munoz et al. 2024), and sperm RNA and epigenetic inheritance of acquired traits (Belleannee 2015; Chen et al. 2016b, 2021; Chu et al. 2019).

2 Dynamic Features of Extracellular Vesicles During Epididymal Sperm Maturation

2.1 General Biological Properties and Compositions of EVs in the Epididymis

The physiochemical structure of an EV defines its basic function—to carry enveloped matters. In this chapter, we use the term epididymal EVs to annotate all sizes of EVs present in the epididymis, with a diameter ranging from 50 nm to several micrometers, e.g., apical blebs from epididymal epithelial cells (Cooper 2005; Cornwall 2009; Sullivan et al. 2007), while the term epididymosomes refer only to small EVs (or exosomes) with a diameter between 50 and 200 nm. The specific functional roles of lipid bilayer–enclosed membranous EVs are also determined by their intrinsic compositions including (1) proteins and/or enzymes on the membranes or in the intraluminal contents, (2) lipids on the membrane envelopes, (3) genetic materials including RNAs and fragmented DNAs, and (4) inorganic elements. EVs are routinely featured by extracellular particles surrounded by lipid bilayer integrated with specific membrane proteins, which can be transmembrane proteins serving as specific makers, such as CD9, CD63 and SNARE as the common markers of epididymosomes (Mear et al. 2024; Thimon et al. 2008; Zhou et al. 2018). EVs may also contain enzymes and contribute to some active catalytic activities, including glutathione-redox-based enzyme activity in epididymosomes (Nixon et al. 2018). In the non-epididymal EVs, other nutrients, for example, carbohydrates such as glycoproteins and glycolipids (Yanez-Mo et al. 2015), inorganic elements such as calcium (Olli et al. 2018), and micronutrients like vitamins, have also been associated with EVs, including vitamin B12-binding proteins in the xenobiotic EVs in the gastrointestinal tract (Juodeikis et al. 2022). In the epididymal EVs, while the primary research has focused on proteins, amino acids, lipids, and RNA contents, research regarding essential micronutrients is still scarce, and further research is required to explore this aspect. For example, lipid-soluble membrane-embedded micronutrients such as vitamins A, D, E, and K have been shown to be important for sperm maturation in the epididymis, as reviewed in Shum et al. (2022), whereas EVs artificially conjugated with vitamin A can exert beneficial physiological functions in somatic stem cells (You et al. 2021). However, endogenous micronutrients and their functions in the epididymal EVs remain to be explored. For updates on the compositions of epididymal EVs, refer to the recent comprehensive review article by Mear et al. (2024).

2.2 A General Veiw of the Origin and Biogenesis of EVs in the Epididymis

In the epididymis, epididymal EVs can be derived from different types of epithelial cells, sperm, and even immune cells that are able to reach the epididymal lumen. Distinctive kinds of epididymal EVs from different cell types in different regions of epididymis allow the exchange of specific constituents of sperm as well as epididymal cells (Frenette et al. 2010; Mear et al. 2024; Sullivan et al. 2007; Thimon et al. 2008; Zhou et al. 2018). As illustrated in Fig. 1, different kinds of epididymal EVs and vesicle-like structures exist on the apical surface of epididymal epithelial cells and sperm in the epididymal lumen. Given that the epididymis connects to the testis, it is also conceivable that the sources of EVs in the epididymis include not only EVs released from epididymal epithelial cells but also testicular EVs that accompany

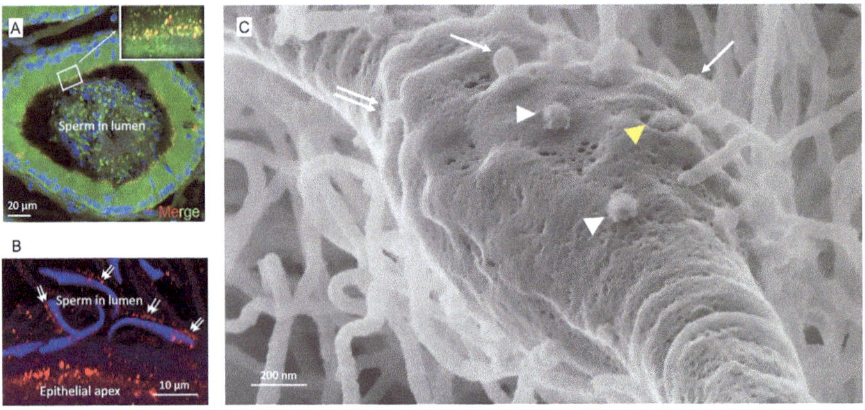

Fig. 1 Various extracellular vesicle-like structures in the epididymis and sperm. (**a**) Vesicle-like structures (yellow), double-labeled for TRPV6 (green) and TMEM16A (red), are present at the apical surface of epithelial cells in the initial segment of a rat epididymis. Inset: higher resolution image showing the vesicle-like granular structures on the stereocilia on the apex of epididymal principal cells, which are in contact with the spermatozoa in the lumen. (**b**) Vesicle-like structures labeled for MGP in red on the sperm heads (in blue labeled with DAPI) in the lumen of rat epididymis. (**c**) Helium ion imaging showing exosome-like membranous structures on the cytoplasmic droplet of a mouse cauda spermatozoon. Distinctive characteristic vesicles are identified on the membrane surface. Some vesicles exhibit a smooth appearance (arrows), while others display rough surfaces (arrowheads). These vesicles are likely internalized by sperm through a phagocytotic process (yellow arrowhead), or via a more complex mechanism involving membrane protrusion budding from the cytoplasmic droplet (double arrow). Images are adapted from (Gao et al. 2016; Ma et al. 2019; Paunescu et al. 2014) with permission and modifications

sperm from upstream testicular contents, such as CD63-positive exosomes (Jankovičová et al. 2024). Spermatozoa are also expected to release exosomes during their transit through epididymis, as shown by the helium ion imaging (Fig. 1c) of some small EVs that can be conceived to be derived from the epididymal spermatozoa. On the other hand, since epididymal tubule is an open tubule, it can easily be invaded by pathological species (Fraczek and Kurpisz 2015; Michel et al. 2016; Pleuger et al. 2020; Rodrigues et al. 2008; Stammler et al. 2015). Consequently, some specific EVs present in the epididymal luminal contents may originate from xenobiotic sources, as has been reported in non-reproductive system under normal or pathophysiological conditions (Juodeikis et al. 2022), although further investigation is still required to verify.

Sperm cytoplasmic droplets in the epididymis. As mentioned above, EVs with various sizes can be derived from distinct origins, mainly from epididymal epithelial cells. In the epididymis, there is another specific type of membranous EVs—the cytoplasmic droplets. During epididymal maturation, one obvious change of spermatozoa is the migration of their cytoplasmic droplets (Cooper 2005), which are absent in human ejaculated sperm (Cooper 2011; Rengan et al. 2012; Xu et al. 2013). Defect in droplet migration in spermatozoa associates with a decrease in sperm fertility. The membranous vesicular cytoplasmic droplets release from sperm during maturation are with diameter greater than 2 μm (Cooper 2011; Hermo et al. 2019; Rengan et al. 2012), which are physically similar to large EVs and can be separated easily by gravitation from epididymosomes, or the small to medium EVs (Wang et al. 2023; Yuan et al. 2013). However, without microscopic or specific biomarker examination, it is still unlikely to clearly distinguish between cytoplasmic droplets and large EVs. In addition to their conventional functions, including nutrient storage, membrane maintenance, signal transduction, and removal of excess cytoplasm, cytoplasmic droplets have been reported to serve as the platform for the shuffling of small RNAs (Wang et al. 2023). This may have been reflected in different species of small EVs, i.e., epididymosomes, which can be distinguished by their physical features on a sperm cytoplasmic droplet at the apical surface of epithelial cells, and some can "fuse" with the plasma membrane of spermatozoa, as shown in the helium ion imaging in Fig. 1c.

Cell-type specific biogenesis of EVs in the epididymal epithelium. Epididymal EVs directly released from the major cell types including principal cells and clear cells through apical secretion have been well documented in rodents and human epididymis (Barrachina et al. 2022; Battistone et al. 2024; Mear et al. 2024; Sullivan 2015; Sullivan et al. 2007; Thimon et al. 2008), including the release of epididymosomes through the process of apocrine secretion of principal cells, a process occurring more frequently in the epididymal proximal region (Cornwall 2009; Hermo and Jacks 2002; Sullivan et al. 2007). Recently, the use of transgenic mouse models has enabled the characterization of cell-type-specific EVs, including those derived from V-ATPase expressing clear cells in the epididymis, which deliver specific epididymal proteins to sperm (Barrachina et al. 2022). It has been reported that immunological monocytes, known as halo cells in the epididymis, can occasionally be present in the luminal side (Robaire and Hinton 2015). Additionally, the dendritic cells (Battistone et al. 2024; Da Silva et al. 2011) are known to be capable extending their dendrites through the epithelium, reaching into the lumen (Niess and Reinecker 2006). Our previous study also found that basal cells in the epididymis can extend antenna-like body projections which traverse the tight junction barrier to reach the lumen (Shum et al. 2008). Although the direct contribution of basal cells toward epididymal EVs still remains to be investigated, the apical-reaching property of basal cells can allow them to function as sensors for non-androgenic luminal hormonal factors (Kim et al. 2015; Shum et al. 2008), in addition to the basolateral hormonal sensor cells (Cheung et al. 2005; Leung et al. 2004). With the apical-reaching property, basal cells can also act as transmitters to communicate their findings to the neighboring proton-pumping V-ATPase-expressing cells to modulate their acid secretion and the fluid secreting principal cells. Through such process, the secretion of epididymal EVs is presumably modulated. Recently, the basal cells in the epididymis have been reported to express primary cilia and safeguard tissue homeostasis (Augiere et al. 2024; Bernet et al. 2018; Girardet et al. 2019). It is conceivable that basal cells may also contribute to the formation of the epididymal EVs mixture by releasing ciliary EVs into the epididymal lumen when they reach beyond the tight-junction during physiological process, as speculated by the

Belleannee's group (Vinay and Belleannee 2022), although this remains to be verified.

Generic Regulation of EVs Biogenesis in the Epididymis. The biogenesis of EVs defines the function of EVs. The biogenesis of EVs, including epididymal EVs, is a complex, highly regulated process that involves several sequential stages, from the initial formation to the eventual release of fully mature EVs into the extracelular environment. In addition, different types of EVs can involve specific formation mechanisms and encompass specific groups of constituents either on their membranes or within their interiors. These cargo molecules dictate targeting specificity and functional roles upon reaching recipient cells. Although the detailed molecular mechanisms of biogenesis of EVs in the epididymis remain to be verified, the current understanding of EVs biogenesis in general can be categorized according to their size (Jeppesen et al. 2023). Small vesicles, or exosomes, are typically derived from the endosomal pathway, and multivesicular bodies (MVBs) and endosomal sorting complex required for transport (ESCRT) are involved in the formation, maturation, and release of these small EVs. Larger size microvesicles, a type of macro-EVs, are formed directly from the plasma membrane via mechanisms including membrane remodeling, caspase activation, and changes in lipid compositions. However, apoptotic bodies are generated during programmed cell death. Some larger EVs with sizes approximately 4 μm in diameter containing damaged mitochondria and protein aggregates can be released in an unknown mechanism but may involve autophagic machinery (Jeppesen et al. 2023). Environmental pH stress can influence the secretion and uptake of CD63-positive EVs (Nakase et al. 2021). A stiff matrix has been shown to induce EVs secretion under tumorigenic tumorous condition (Zhang et al. 2018). Additionally, EVs can be affected by pathogenic infection (Zhang et al. 2018) and can be negatively regulated by inhibitory interventions (Catalano and O'Driscoll 2020). However, the underlying mechanisms of these regulatory processes remain largely unclear.

In the epididymis, it has been reported that an unusual secretory mechanism exists for epididymosomes, involving the apocrine secretion of epididymosomes by epididymal principal cells (Hermo and Jacks 2002; Sullivan 2008; Sullivan and Saez 2013). In apocrine secretion process, small EVs are enclosed in a big bleb and shed into the epididymal lumen. These apocrine blebs are predominantly released from the principal cells in the caput epididymidal region and transfer proteins and essential components to spermatozoa in a saturable, time- and temperature-dependent manner; they are more efficient at acidic pH 6.5 and zinc-sufficient conditions (Sullivan 2008). Interestingly, receptor-ligand proteins (e.g., RABs and SNAREs) known to be involved in vesicle trafficking and fusion were found in different populations of epididymal EVs (Skerget et al. 2015; Sullivan and Saez 2013; Thimon et al. 2008; Zhou et al. 2019). These proteins may participate in the cargo exchanges in spermatozoa; one may raise an open question that whether biogenesis of EVs can occur in situ in the viscous fluid of the epididymal lumen. As demonstrated in Fig. 1, EVs can be derived from different types of epithelial cells in the different regions of epididymis, as well as from the spermatozoa in the lumen. Indeed, the characteristics and contents of EVs depend on the sequential processes of biogenesis and exchange between epithelial cells and sperm in the lumen. Although the exact mechanisms of the biogenesis of epididymal EVs have not been fully elucidated, several putative mechanisms of exchange between EVs and sperm have been described. The first mechanism involves the direct exchange of proteins via lipid rafts, which act as a platform to facilitate the fusion of the epididymal EVs and sperm membranes (Girouard et al. 2009; Griffiths et al. 2008; Sullivan and Saez 2013). The second mechanism proposes that proteins are transported through the formation of a transient fusion pore between the epididymal EVs and the underlying sperm plasma membrane, as implied in the high-resolution helium-ion imaging in Fig. 1c. It has been speculated that the interaction between the EVs and the spermatozoa may not involve complete fusion of the respective membranes (Zhou et al. 2019). Instead, a transient fusion pore may form upon attachment of an epididymal EV, allowing material transfer; subsequently, the EV is released once the transfer is complete. The third mechanism involves the delivery of glycosylsphatidylinositl (GPI)-linked proteins by the epididymal EV to the outer lipid layer of the sperm surface (Griffiths et al. 2008; Patel et al. 2013). The epididymal EV is delivered by CD9, a tetraspanin responsible for creating a fusion active site on the membrane. This faciliated fusion mechanims deliver transmembrane proteins including plasma membrane Ca^{2+}-ATPase 4 (PMCA4) to the sperm surface (Martin-DeLeon 2015). In addition, it has bee reported that Arrdc4-dependent EV biogenesis is required during sperm maturation (Foot et al. 2021) and MFGE8 is implicated in mediating mouse sperm-EV interactions (Trigg et al. 2021). Moreover, SYPL1 was found to be involved in the vesicular pathway that defines sperm cytoplasmic droplet formation (Liu et al. 2023).

2.3 Generic Functions of EVs in the Epididymis

In spermatozoa, the replacement of histones with protamines lead to a high degree of DNA condensation, rendering the classical processes of transcription and translation intrinsically unfeasible. Sperm cells are therefore completely dependent on their extracellular eivironment for survival, functional maturation changes and protectin against oxidative stress, expecially during transit through the epididymis where they achieve functional maturity and meet metabolic demands (Cornwall 2009; Gao et al. 2018; Gervasi and Visconti 2017; Robaire and Hinton 2015). Therefore, epididymal EVs contribute to these primary functions of metabolism and maturation of sperm in the epididymis. Different sources and types of EVs can carry specific membrane surface proteins and endogeneous components, thereby mediating specific functions and promoting the remodeling of spermatozoa during their maturation processes in the epididymis (Gervasi and Visconti 2017; Martin-DeLeon 2015; Mear et al. 2024; Nixon et al. 2018; Skerrett-Byrne et al. 2022; Sullivan 2015; Sullivan et al. 2007, 2019; Sullivan and Mieusset 2016; Thimon et al. 2008).

The physiological contributions of EVs to epididymal sperm maturation can be achieved by directly changing sperm morphology, sperm function, and sperm composition, or indirectly changing the compositions and physiochemical properties of the luminal fluid or affecting the function of epithelial cells. It is attributed to the progress of proteomic analysis of epididymal contents, which has greatly helped to clarify the potential contribution of epididymal EVs to the functional transformation of sperm (Barrachina et al. 2022; Fazzio et al. 2024; Martin-DeLeon 2015; Nixon et al. 2018; Skerrett-Byrne et al. 2022; Sullivan 2015; Sullivan et al. 2007, 2019; Sullivan and Mieusset 2016; Thimon et al. 2008). For the major candidate proteins and their related functions, refer to the recent comprehensive review article by Mear et al. (2024). The redox stress protection also involves EVs in the epididymis. For example, several antioxidants have been identified in the contents of epididymal EVs, such as glutathione S-transferase (GST) (Sullivan et al. 2007), selenium-independent epididymis-restricted antioxidant glutathione peroxidase 5 (GPX5) (Taylor et al. 2013), biliverdin reductase A (BLVRA) (Sullivan 2015), and the potential dead sperm marker epididymal sperm binding protein 1 (ELSPBP1) (Barrachina et al. 2022). During maturation, sperm cells dramtically remodel their plasma membrane and there are correlations between changes in the epididymal lipid compositions and sperm membrane lipids (Bjorkgren et al. 2015; Girouard et al. 2011; Huszar et al. 1997; Rejraji et al. 2006). For example, it has been reported that the lipid rafts play key roles in sperm maturation, and raft-like structures on the membrane of EVs are thought to be the site of signaling complex (Sullivan et al. 2007; Zhou et al. 2019) and that epididymosomal EVs contribute to the clearance of proteins from the lumen by binding and delivering proteins to cells for endocytosis (Cornwall 2009). EVs can also contribute to the specific physiological features of epididymal fluid and nutrition balance (Gatti et al. 2004; Zhou et al. 2018). For example, the EVs present in the epididymal fluids can directly modulate its features. This modulation occurs through enzymatic activities associated with the EV membrane, such as the lactase, which breaks the sugar lactose (Grimalt et al. 2000), and hyaluronidase, which breaks down hyaluronic acids (Martin-DeLeon 2006). The effects of EVs on epithelial cells have been demonstrated under in vitro conditions, supporting the notion that EV-mediated delivery of sncRNA cargo to epididymal epithelial cells could functionally modulate gene expression patterns in these cells in vivo in a paracrine manner (Belleannee et al. 2013), although more studies are still needed to elucidate the underlying mechanism. Nevertheless, these studies clearly demonstrate the importance of the influence of the epididymal EVs on sperm function, which may affect the subsequent epididymal-dependent fertilization potential of sperm, and even the reproductive outcome and the health of offspring, as mentioned in the review articles (Fazzio et al. 2024; James et al. 2020; Mear et al. 2024).

3 Exploring the Role of Extracellular Vesicles in Redox Metabolism in the Acidic and Decalcitonic Epididymal Microenvironment

3.1 Redox Stress and Redox Coupling Network in the Epididymis

Roles of redox stress in the epididymis: a friend or a foe or an energy metabolism balancer? In an oxygenated environment, redox reactions are considered to be the central reactions of life and are intrinsically linked to energy metabolism (Feelisch et al. 2022; Lennicke and Cochemé 2021; Santolini et al. 2019; Sies et al. 2024; Zhang et al. 2024). A redox reaction, or oxidation-reduction reaction, is a process of electron exchange between pairs of chemical species, so all biomolecules or bio-compounds can engage in "redox" (i.e., electron transfer) processes because their atomic constituents can either act as electron donors (reductants) or acceptors (oxidants). Redox reactions are used in basic cellular respiration to generate energy (e.g., ATP) in aerobic cells and to utilize nutrients to build essential cellular components, whereas at physiological low levels of beneficial redox stress (also known as redox eustress), stress is used for redox signaling and redox metabolism. Redox stress refers to an imbalance between the production of reactive oxidative species (ROS, a term used in this chapter collectively to include the free radical-derived species, as well as the reactive nitrogen species, RNS, and reactive sulfur species, RSS) and the ability of a biological system to detoxify them or repair the resulting damage. For the definition of more of these redox-related terms, see Fig. 2. In case of sperm, redox stress can indeed promote their respiration and motility in mice and men (Moon et al. 2024). However, redox metabolic imbalance can also lead to deviations from physiological redox eustress. When excessive oxidative stress or reductive stress exceeds the ability of cells to repair the redox state of biomolecules, cytotoxic effects can occur, leading to life-threatening consequences. To sustain life, almost all cellular living beings have been successively generating, transforming, and eliminating the free radicals and ROS in a variety of cell processes, and a series of functional and localization-specific cellular mechanisms are required to maintain redox balance. Thus, redox metabolism involves finely regulated network of redox reactions and their cellular respiratory outputs. Although a high prevalence of sperm DNA oxidative damage was found in infertile men (Aitken et al. 2014; Oleszczuk et al. 2013), in fact only about 40% of infertile men showed abnormally increased ROS levels (Agarwal et al. 2014), and the association between ROS and sperm impairment and male infertility may not be a direct causal relationship. In the epididymis, the presence of small amount of ROS specifically is not only the result of normal physiological processes but also necessary for sperm functional maturation. It is known that epididymal spermatozoa and epithelial cells can generate ROS as by-products in various metabolic mechanisms and ROS themselves participate in signaling and functional processes together with several specific antioxidant systems that are abundant in epididymal cells and sperm (Aitken and

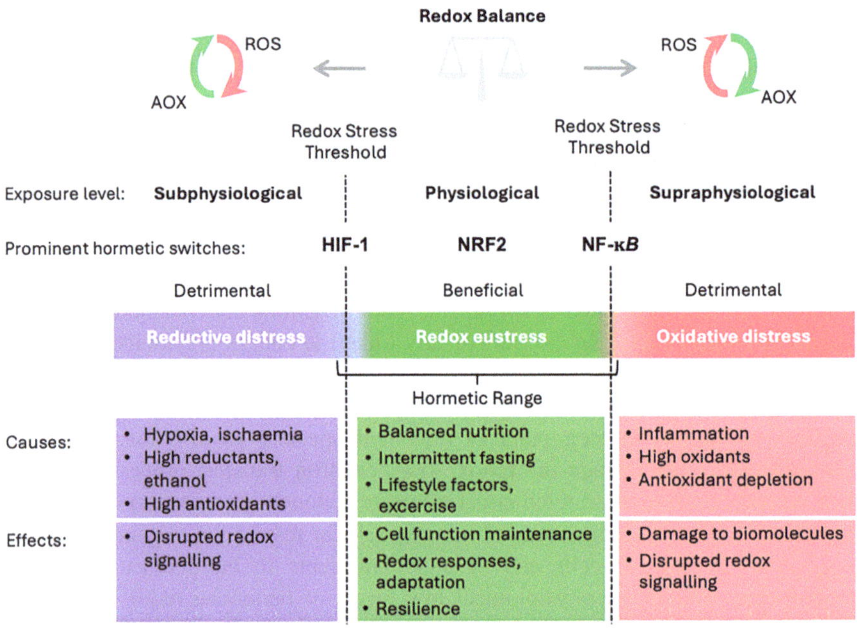

Fig. 2 An overview of the functional role of redox stress in biological systems. Many free radicals (a common form of reactive oxidative species (ROS) in the body, which also collectively includes reactive nitrogen species and reactive sulfur species) are unstable and highly reactive chemical species. They can either accept an electron from or donate an electron to other molecules, thereby behaving as oxidants or reductants, respectively. When the ROS and antioxidants (AOX, a term usually including the mixed meaning of both antioxidants and anti-reductants) are in balance, a redox eustress condition is in a state-of-qua and physiological processes are maintained in a healthy state. Antioxidants inhibit oxidation by donating electron(s) to radicals, thereby preventing oxidative distress, a condition characterized by excessive ROS beyond physiological levels. In contrast, anti-reductants act oppositely in redox reactions; they inhibit reduction by accepting electron (s) from radicals, thus preventing reductive distress, which occurs when ROS level fall below their physiological threshold. These effectors (e.g., AOX) are also considered as secondary redox stress species and can also be generated after interacting with the active primary species, e.g., free radicals directly released from enzymatic processes. (Illustration, modified from Manford et al and Sies et al (Manford et al. 2021; Sies et al. 2024), annotates the general understanding of redox balance and its associated physiological and pathophysiological conditions in biological systems, including the epididymal sperm maturation in the male reproductive system)

Vernet 1998; Juarez-Rojas et al. 2021; O'Flaherty 2019; Vernet et al. 2004). For example, two mechanisms have been found in sperm during maturation in epididymis, one from mitochondria via cellular metabolism and the other at the plasma membrane (Vernet et al. 2001) via NADPH oxidase (Musset et al. 2012), which catalyzes the oxidation of NADPH, transferring electrons to molecular oxygen (O_2) to generate superoxide anion ($O_2^{\bullet-}$) and transfers ROS superoxide anions ($O_2^{\bullet-}$) to superoxide dismutase (SOD), which then converts the superoxide anions into hydrogen peroxide (H_2O_2) (Juarez-Rojas et al. 2021; Vernet et al. 2004). Epididymal spermatozoa then use redox-labile H_2O_2 to form disulfide bridges in the DNA-bound protamines at the nucleus for the final stage of chromatin compaction (Ijiri et al. 2013; Oliva 2006; Pfeifer et al. 2001). These protamine-dependent sperm DNA condensations increase as sperm transit through the epididymis and can be deformed in vitro by reducing reagents (Cabrillana et al. 2017; Ward 2009). Defective compaction of chromatin with protamine disulfide bridges during epididymal sperm maturation is known to lead to impaired sperm motility and infertility, suggesting that ROS has a fundamental physiological role on spermatozoa (Juarez-Rojas et al. 2021; Oliva 2006; Ward 2009).

Redox-active signaling molecules in cells play essential roles in physiological processes, including hydrogen peroxide (H_2S) and nitric oxide (NO), in addition to H_2O_2, which plays a functional role in sperm chromatin condensation as mentioned above. These redox-active gases have been reported to function in the epididymis via specific intracellular and intercellular mechanisms. The nitrous-based oxidative active species, NO, is a free radical and well known as EDRF (endothelial-dependent relaxation factor). In the epididymis, our previous study has found that NO may participate in the intercellular communication between the angiotensin-2 type II receptor AGTR2-expressing basal cells and the proton-secreting clear cells to promote proton secretion and luminal acidification (Shum et al. 2008). This intercellular interplay involves the activation of basal cells by ANGII from the luminal side, which can extend their body projections through the tight junction to sense luminal hormone ANGII, and can act on the adjacent clear cells, possibly through the release of the downstream effector NO as the signaling molecule (Siragy and Carey 1997). The NO then acts locally on the soluble guanylyl cyclase which is enriched on the paracellular plasma membrane of adjacent clear cells. This interaction subsequently triggers the insertion of proton-pump V-ATPase into the apical membrane of clear cells via an actin remodeling mechanism, thereby promoting acid secretion (Shum et al. 2008, 2011a). Although NO itself is not considered as one of the primary oxidative stress sources in cells, however, when it reacts with $O_2^{\bullet-}$, which can be generated from NADPH oxidase, oxidoreductases, uncoupled NO synthase, and mitochondrial respiration, a highly reactive and damaging molecule peroxynitrite ($ONOO^-$) can be formed. Peroxynitrite is a potent and short-lived oxidant and can oxidize lipids, proteins, and DNA, leading to deleterious cellular oxidative stress. H_2S is a sulfide-based redox signaling molecule; its physiological function has drawn the attention of researchers in the recent years (Cirino et al. 2023). H_2S is now considered as at least one of the "EDHFs" (endothelial-dependent hyperpolarization factors), if not the only one, and known as gasotransmitter. Many protective effects of H_2S have been reported, and it can be used to treat male reproductive system diseases (Song et al. 2024). For example, it has been shown to promote epididymal transepithelial transport of K^+ ions into the lumen of rat cauda epididymidis by involving ATP-sensitive K^+- channels and Ca^{2+}-activated K^+-channels (Gao et al. 2018), playing a role in maintaining a high potassium-to-sodium ratio for sperm maturation (Levine and Marsh 1971). It is also conceived that H_2S can affect sperm motility in a bell-shaped concentration-response manner (Cirino et al. 2023). That is, in a slow releasing mode, H_2S improves motility of asthenozoopermic sperm in men, but a fast H_2S-relase donor decreases sperm motility in mouse and boar semen; however, under oxidative stress conditions, H_2S donation protected against the loss of spermatozoa motility (Cirino et al. 2023). Excessive levels of H_2S are hazardous and H_2S toxicity is attributed to glutathione (GSH) depletion in hepatocytes, which can occur even in acidic condition and may involve a divalent-dependent iron (Fe) activation and ROS release (Truong et al. 2006). The deleterious effects and mechanisms due to imbalanced, mislocated, or excessive H_2S generation in epididymal cells during sperm maturation in the unique luminal microenvironment remain to be elucidated.

In the epididymis, a complex intercellular interplay is involved in the luminal microenvironment. The different cell types of epididymal epithelium also have their own specific set of genetic composition and expression, and thus specific cell types have their own respiration activities and redox stress response system during sperm maturation. Epididymal epithelial cells are comprised of at least four cell types, including nutrient-transporting principal cells, acid-secreting clear cells or narrow cells, the basal cells that usually sit beneath the principal cells or clear cells occasionally sending their narrow projection to reach out to the luminal side, and some immunological cells from immune system (Breton et al. 2019; Da Silva et al. 2011; Elfgen et al. 2018; Robaire and Hinton 2015; Shum et al. 2011b, 2022). For example, it is known that in the epididymis, clear cells express high levels of proton pump vacuolar ATPase (Breton et al. 1996; Brown and Breton 1996) and have obviously greater amount of mitochondria than the adjacent principal cells, whose role is mainly to transport solutes (Leung et al. 2004), and the underlying basal cells can function as luminal hormonal sensors and intercellular communicator (Shum et al. 2008). Recently, single-cell transcriptomic studies revealed that the genes involved in mitochondrial energy production are enriched in the mouse epididymal principal cells and clear cells but are depleted in basal cells and other peripheral non-epithelial cells (Rinaldi et al. 2020), suggesting that epididymal epithelial cells exhibit a higher level of cellular respiration metabolism, which play an essential role in acidifying the epididymal lumen (Breton and Brown 2013; Breton et al. 1996; Park et al. 2017; Pastor-Soler et al. 2005; Shum et al. 2008). Mitochondrial transcripts are also more abundant in the corpus and cauda epididymidis than the proximal regions (Shi et al. 2021), consistent with higher number of proton-pumping epithelial clear cells in these regions (Shum et al. 2009, 2011b). The distal epididymal region is the main site where sperm are stored in dormant stage and is occasionally invaded by pathogenic exogenous agents, thus requiring a complex protection against ROS to prevent the pathogenic conditions of epididymitis (Stammler et al. 2015; Weigel Munoz et al. 2024). It has been reported that lower temperature in epididymis facilitates the sperm storage by enhancing oxygen availability (Djakiew and Cardullo 1986). Although the mechanisms is still unclear, higher mitochondrial activity in both the epithelial cells of the epididymal epithelium and the sperm cells within the lumen may contribute to the evolution of the male reproductive ductal system outside the body. This is supported by evidence that the temperature of mitochondria can reach up to 50 °C, which is at least 10 °C higher than the ambient temperature surrounding the cells (Chrétien et al. 2018). Mitochondria are intricately linked to redox metabolism and have huge impact on sperm function. Sperm mitochondria dysfunction, such as abnormal production of ROS and thermal energy, is known to be associated with male infertility (Durairajanayagam et al. 2021; Gibb et al. 2024; Guthrie et al. 2011; Sun et al. 2014b; Vijayaraghavan and Hoskins 1990). Therefore, the complex intercellular interplay of epididymal epithelial cells and sperm is crucial in maintaining epididymal homeodynamics to ensure sperm function and reproductive health.

The regulation of integrative cellular stress responses remains a critical biological question concerning when and how the various cellular stress responses are activeted following redox imbalance. Thus far, the conventional key players of redox-regulators are hypoxia-inducible factor (HIF-1), nuclear factor erythroid 2-related factor 2 (NRF2), and nuclear factor kappa-B (NF-κB) (Fig. 2). The physiological roles of these redox regulators have also been reported in the epididymis. In other biological systems, these redox regulators are known to operate together at low basal levels and integrate with stress responses to maintain redox balance under physiological conditions (Sies et al. 2024). It is known that HIF-1 is regulated by hypoxic and non-hypoxic activators and responds to chronic acidity, resulting in the generation of pro-oxidative bio-compounds and thus eliciting or resetting the cell-oxidative stress; thus, HIF-1 is considered the redox-regulator between hypoxia and normoxia conditions (Sies et al. 2024). In the rat epididymis under normoxic conditions, both isoforms of HIF-1 are found to be expressed only in the principal cells. Isoform HIF-1a has higher levels in the middle to distal regions, whereas isoform HIF-1b is uniformly distributed in the principal cells of all regions, suggesting its role in maintaining redox homeostasis to protect sperm from hypoxia by principal cells (Palladino et al. 2004). NRF2 has been implicated as the redox balancer in the normoxia and oxidative conditions and a key transcription factor for antioxidant response elements and cell detoxification (Sies et al. 2024). In the cardiovascular system, hyper-reductive stress induced by constitutive expression of NRF2 can lead to pathological conditions and significant mortality (Shanmugam et al. 2020; Xiao and Loscalzo 2020), whereas chronic pro-reductive stress induced by the organ-specific expression of NRF2 favors the fight against oxidative stress (Jyothidasan et al. 2022). This is consistent with the fact that the epididymal lumen is acidic and that NRF2 expression is higher in epididymis than in testis (Wajda et al. 2016), and NRF2 genetic deletion can lead to lipid peroxidation in sperm and a decrease in the expression of antioxidants GST and SOD2 (Valipour et al. 2024). Another redox regulator, NF-κB, which functions together with ERT (endoplasmic reticulum stress) pathway and STAT (signal transducer and activator of transcription) pathways, was found in the rat initial segment to prevent apoptosis under the regulation of testicular lumicrine factor (Xu et al. 2011), suggesting that the oxidative stress is high in this region of epididymis. Hypoxia is a hallmark of metabolic disorders with a decline of oxygen level in the tissues that induces hypoxic responses. Hypoxia-induced responses trigger the release of EVs, which participate in complex intercellular communication, while failure to adapt to prolonged or severe hypoxia leads to cell death (Bister et al. 2020). It has been reported that hypoxia-sensitive cells trigger the activation of natural killer cells that contribute to impairment in the rat epididymal sperm via the KLF4-ASH1L-ICAM-1 axis (Zeng et al. 2023), and natural killer cells–derived EVs serve as immune sensors (Federici et al. 2020). Hypoxia is known to be associated with male infertility (Cornejo-Guerra et al. 2024; Li et al. 2021). However, how hypoxia-induced responses of EVs differ from those under normoxic conditions and their relationship to male infertility remain an open question.

Excessive free radicals and imbalanced ROS have destructive effects on epididymal sperm. Sperm are highly susceptible to oxidative active species due to the special composition of their plasma membrane that is rich in polyunsaturated fatty acids (PUFAs) and their special cytologic and metabolic features that lack cytoplasm to mount stress responses. While some basic functions and metabolic activities require a small amount of ROS in epididymal epithelial cells and sperm, chronic exposure to excessive oxidative stress, including harmful levels of ROS and H_2O_2, exerting deleterious effect on sperm and epididymal function is well documented (Agarwal et al. 2020; Aitken and Vernet 1998; Juarez-Rojas et al. 2021; O'Flaherty 2019; Vernet et al. 2004). These chemically dissimilar species can be the primary oxidative free radicals, which carry unpaired electrons and react intravenously with other free radicals or non-radicals as the secondary reactive species. Examples of secondary reactive species are the oxyradicals, including $O_2^{\bullet -}$ and hydroxyl radical ($^{\bullet}OH$), and the free radical precursor H_2O_2. These oxidative active species deform many enzymes, thereby decreasing their affinity for their substrates or coenzymes and their activities, leading to detrimental consequences, such as loss of protein function due to oxidative chemical modifications, adduction of aldehydes produced from lipid peroxidation, and altered function of membrane proteins. The pH sensitivity of free radicals that can form chelating complex with calcium ions (Ca^{2+}) or other free ionic nutrients is an important consideration, as the pH can significantly affect the formation and stability of these organic complexes (Carter 1995; Hamamah and Gatti 1998). Elevated pH increases hydroxyl ion (OH^-) concentration, which promotes the precipitation of Ca^{2+}-containing compounds by shifting solubility equilibria. This process may lead to pathological calcification in tissues such as blood vessels or kidneys (Gorlach et al. 2015; Shum et al. 2022). Consequently, the availability of free Ca^{2+} for chelation by radicals and other chelating species, such as redox-dependent GGCX (γ-glutamyl carboxylase) carboxylation of MGP (matrix Gla protein) in calcium homeostasis, can be limited. Notably, this process is also subjected to the regulation of extracellular pH and divalent ions, as demonstrated in our previous studies (Gao et al. 2016; Ma et al. 2019). As a general working principle, a balanced presence of both antioxidants and anti-reductants appears to be beneficial to promote health and minimize adverse reactions, whereas most biomolecules are multifunctional, especially in vivo. While extensive proof is still needed, the authors here hypothesize that a pro-reductive environment constitutes a fundamental defense mechanism enabling the epididymis to withstand oxidative stress in this oxygen filled world. This pro-reductive condition is proposed to facilitate the passage of fragile gametic sperm through the epididymis by adapting a specific redox signaling environment, nediated by cooperations among spermatozoa, various types of epithelial cells, and, occasionally, immune cells reaching to the lumen. More evidence from the literature are discussed in the subsequent sections.

EVs are generated in response to stress. Among the various stress conditions, redox stress occurs due to imbalances in free radicals and aberrant levels of signaling molecules, such as H_2O_2, H_2S, ROS, free radicals, as well as hypoxia factors. The redox-responsive EVs then act as intercellular communicators and microenvironmental signal transducers and convey redox signals to the recipient cells (Chiaradia et al. 2021; Chu et al. 2020; Gai et al. 2024; Jeppesen et al. 2023; Yanez-Mo et al. 2015). These redox-responsive EVs even have been reported to be associated with transmitting stress signals across generations (Chan et al. 2020; Conine et al. 2018; Sharma et al. 2016, 2018). In the non-reproductive system, such as in the cardiovascular system, the contents carried by EVs can alter the redox status; for example, heme and iron (Fe) in the EVs can cause oxidative damage to phospholipids on cell membrane and scavenge NO free radicals, thus interrupting the NO-mediated signaling function in maintaining the homeostatic regulation of vascular tone (Chatterjee et al. 2020). In the neuronal system, it has been reported that amyloid-β-induced ROS increase neuron-derived EVs carrying toxic tau (Fowler et al. 2024). In kidney—an excretory organ with epithelial tissue that functions in many ways similar to the epididymis—EVs are involved in the metabolic reprogramming and redox signaling in chronic pathological conditions (Aparicio-Trejo et al. 2022). In the epididymis, although significant progress has been made, most existing literature focus on the functional molecular relationships and mechanisms, while research on perception, transformation, and elimination of epididymal EVs, especially those related to redox metabolism, is still in their infancy and needs further exploration (Ali et al. 2023; Baskaran et al. 2020b; Mear et al. 2024; Paul et al. 2021).

A complex redox coupling network is present in the epididymis. The underlying regulatory mechanisms of this epididymal redox network are emerging, yet much remains to be revealed (Fig. 3). During epididymal sperm maturation, the main cellular redox couples mainly include the nicotinamide adenine dinucleotide (NAD^+) and reduced NADH pair, the phosphorylated NAD^+ ($NADP^+$) and reduced $NADP^+$ (NADPH) pair, and the reduced glutathione (GSH) and its disulfide (GSSG) pair, which are responsible for the bulk of cellular electron transfer and serve as redox buffers in the epididymal cells and spermatozoa (Juarez-Rojas et al. 2021; O'Flaherty 2019; Robaire and Hinton 2015; Vernet et al. 2004). NADH and NADPH are mainly generated by mitochondria respiration and the pentose phosphate pathway, respectively, and their reductive power drives the reduction of oxidative modification processes and maintains the redox state of hundreds of different proteins and small molecules. The reduction of NADH drives the generation of ATP via OXPHOS (oxidative phosphorylation), the production of lactic acid from pyruvate, and the desaturation of PUFAs, all of which are important factors for sperm maturation. The reduction of NADPH plays an essential role in antioxidant defense and regulates cellular signaling via NADPH oxidases (NOXs), an essential signaling in sperm activities (Cui et al. 2024; Musset et al. 2012; Vermot et al. 2021).

The link between NADPH and the primary targets of cellular redox modifications, the cellular protein thiols, is mediated by the thioredoxins (TXNs) and GSH system, catalyzed by thioredoxin reductases (TRDs) and glutathione reductase (GR,

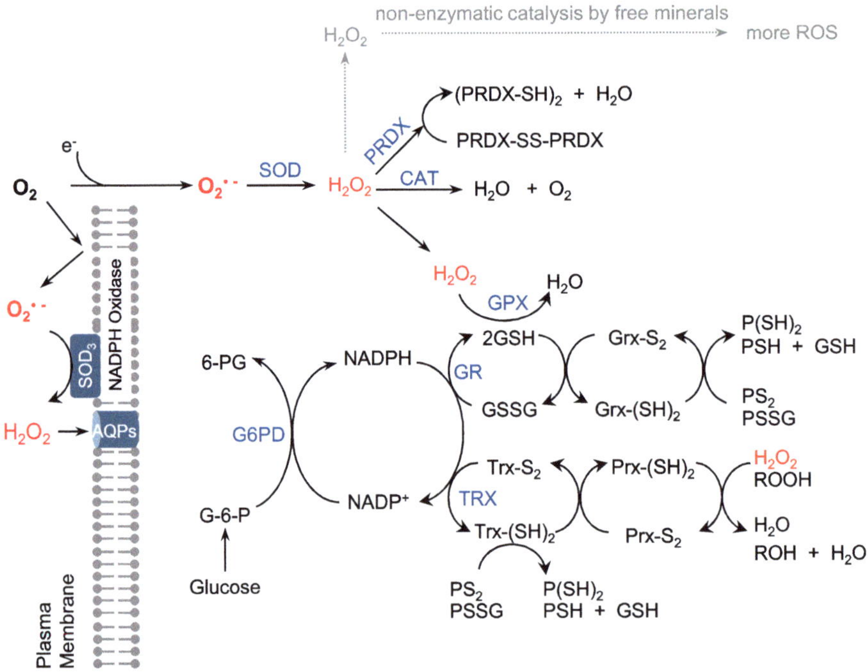

Fig. 3 The cellular redox network in the epididymis. In the epididymis, simialr to the redox network of other cellular systems, the ROS pro-oxidants, which are highly reactive and can induce oxidative modifications of organic matters with dangerous consequences for cell integrity and function, are removed by different antioxidants and enzymes. For example, free radical superoxide anion ($O_2^{\bullet -}$), generated from oxygen molecule, is removed by superoxide dismutase (SOD) by converting it into hydrogen peroxide (H_2O_2), a more stable ROS. On the plasma membrane of sperm, superoxide radical $O_2^{\bullet -}$ can also be generated by NADPH oxidase (NOX5) but can be removed by a specific form of extracellular SOD, SOD_3. Catalase (CAT), restricted to peroxisomes, peroxiredoxins (PRDXs), and glutathione perxidases (GPXs) are also capable of removing H_2O_2 by different reactions that produce H_2O and other products such as O_2. Other thiol-assocaited emzymes, including glutathione reductase (GR) and thioredoxin (TRX), as well as non-enzymatic bioactive substances (i.e., GSH, vitamins and minerals) are also involved in the subsequent dissipation of ROS-related reactions. These enzymes function in accordance with the redox reactions and the properties of redox pairs. GPX oxidizes two molecules of GSH to GSSG. GR catalyzes the recycling of GSSG to GSH and utilizes NADPH as an elecron donor. GSH can also be used by the oxidized glutaredoxins ($Grx-S_2$) to its reduced form $Grx-(SH)_2$, which reduces protein intra-/inter-disulfide (PS_2 or PSSG) into reduced cysteiny residues. The members of the antioxidant network machinery modulate the redox status of the cell by scavenging radical species and repairing (or removing) oxidatively damaged molecules. Non-enzymatic scavenging of pro-oxidants may involve vitamins A, C, and E. However, non-enzymatic dissipation of pro-oxidants by minerals (e.g., Cu, Fe, and Mn) results in increased ROS and can lead to uncontrolled oxidative damage. Image is adapted and modified by referring to citations detailed in the text (Asensi-Fabado and Munne-Bosch 2010; Juarez-Rojas et al. 2021; O'Flaherty 2019; Robaire and Hinton 2015; Seligman et al. 2005; van de Lagemaat et al. 2019; Vernet et al. 2004; Xiao and Loscalzo 2020). Other players in this network also include 6-PG (6-phosphogluconate), G-6-P (glucose-6-phosphate), G6PD (glucose-6-phosphate dehydrogenase), AQPs (H_2O_2 permeable aquaporins), NADPH (reduced $NADP^+$)

also known as GSH disulfide reductase, GSR), respectively (Xiao and Loscalzo 2020). The precursor of NAD^+/NADH pair also plays a major role in spermatogenesis by inhibiting ferroptosis in the testis, but its role in epididymis remains unclear (Feng et al. 2024). These main redox couples work together with other redox-sensitive proteins, including cysteine (Cys) and thiols groups containing compounds, which function in smaller buffering capacity for redox stress not as prominent as the above-mentioned redox couples (Esworthy et al. 2022; Gonzalez et al. 2021). These redox couples also act through other redox-regulating enzymes to maintain the redox balance in the male excurrent duct, including epididymal superoxide dismutase (SODs), catalase (CAT), glutathione peroxidases (GPXs), glutathione-S-transferases (GSTs), gamma-glutamyl-transpeptidases (GGTs), indolamine 2,3-dioxygenase (IDO), and glutathione peroxiredoxins (PRDXs) (O'Flaherty 2019; Robaire and Hinton 2015; Seligman et al. 2005; Vernet et al. 2004).

Non-enzymatic compounds are also involved in the complex redox network in the epididymis. These include vitamins and some necessary elements such as zinc (Zn) and iron (Fe), which can function as antioxidant factors, and they also are integral parts of the epididymal redox-coupling system. The ability of redox-active elements in biological systems, such as iron (Fe), copper (Cu), selenium (Se), chromium (Cr), and cobalt (Co) in redox-balance, to scavenge or generate free radicals and ROS in specific microenvironment has been studied (Mirnamniha et al. 2019; Tvrda et al. 2015). For example, CAT enzyme contains heme and is thus an Fe-dependent enzyme, whereas the functionality of SOD enzyme is mainly contributed by Cu. It is known that these elements can be beneficial or toxic for cells, including spermatozoa, dependent on how they can affect the redox state of the cell (Jomova and Valko 2011). Abnormal changes of these redox-active elemental concentrations of epididymal spermatozoa are found to be associated with male infertility (Zhang et al. 2023). Most of these biomolecules or bioelements are multifunctional and their precise cell-type specific and subcellular spatiotemporal compartments are essential for determining their oxidizing (electron acceptor) and reducing (electron donor) functions. In the epididymis, although the subcellular localization of these redox reactive couplers and scavengers has not yet been fully elucidated, a balanced distribution of both anti-oxidants and anti-reductants within specific cell types and subcellular spatiotemporal compartments appears to be advantageous for promoting health and minimizing adverse reactions, especially in vivo (Baskaran et al. 2020a; Gibb et al. 2020; Juarez-Rojas et al. 2021; O'Flaherty 2019). Recent studies on single-cell transcriptomics analyses provide some insightful information in this regard (Rinaldi et al. 2020; Shi et al. 2021).

Through this network involving both enzymatic and non-enzymatic players (Fig. 3), epididymal cellular redox homeostasis is maintained by reducing equivalent quantities of NADH, NADPH, and GSH, thereby counteracting the oxidative stress and preserving the redox equilibrium within the epididymal lumen, which subsequently ensures sperm homeostasis. Conversely, harmful conditions may arise when the of these reduced molecules become either excessively high or low, leading to either reductive stress or oxidative stress, as implicated in certain pathological processes (Moon et al. 2024; Moustakli et al. 2024; Panner Selvam et al. 2020; Sengupta et al. 2022).

3.2 Acidic Microenvironment Is a Pro-reductive Condition That Sets the Basic Equilibrium for the Redox Coupling Network in the Epididymis

Epididymal acidic microenvironment favors a reductive state. During sperm maturation in the epididymis, luminal acidification is essential, a process that increases the concentration of protons (H^+ ions, also commonly called positive electrons) in the epididymal lumen (Au and Wong 1980; Battistone et al. 2019; Breton et al. 1996; Breton and Brown 2013; Carr et al. 1985; Park et al. 2017; Pastor-Soler et al. 2005; Shum et al. 2008, 2009). The pH of luminal fluid in different regions of epididymis is known to range from 6.5 to 7.2, depending on the species, and in the tail region of epididymis where sperm are primarily stored, the pH is typically around 6.5 to 6.8 (Da Silva et al. 2007; Levine and Marsh 1971; Liu et al. 2022; Shum et al. 2022). The redox reaction, i.e., oxidation-reduction activity, of a biomolecule depends on the chemical properties of the set of compounds or molecules with which it interacts, that is, its redox environment (Santolini et al. 2019; Sies et al. 2022; Sies et al. 2024). Redox reaction is carried out by two half-reactions, oxidation and reduction, and the balance of redox reaction can be affected by the pH conditions, i.e., the acidic protons (Fig. 4). Therefore, for reactions that require hydrogen or hydroxyl ions to balance the redox equation, an acidic microenvironment promotes a pro-reductive condition, i.e., a state that favors equilibrium towards the reductant side (Carter 1995). Acidity promotes reduction by facilitating electron transfer and maintains a redox reaction pairs at the reducing state. Thereby, acidity can play a role in promoting reduction reactions and maintaining the reducing state by providing protons and stabilizing reduced species. Thus, it can be inferred that the acidic microenvironment in the epididymis is a pro-reductive conditon that favors reduction or maintains the state or the lifespan of the reductant in a given redox reaction. It is known that physiological reductants such as GSH and vitamins are essential for health, and such reduction-promoting conditions in the epididymis do favor the reduced form of these reductants. However, whether this is universally beneficial under all physiological conditions depends on the equilibrium balance of the reductants and the environment in which are encounted. Excessive or imbalanced levels of reductants, or their presence in inappropriate contexts, can indeed be harmful. For example, overproduction of reactants in extreme scanarios may lead to the generation of reduced species, either due to excess proton donation from acids or under deleterious reduction potentials, such as hypoxia induced by thiol reductive stress or thiol toxicity (Ravi et al. 2023), which drive harmful reduction reactions and the occurrence of hyper-reductive stress (Sadeghi et al. 2023). Paradoxically, it is also known that acidity can increase oxidizing properties of the disulfide bond-containing

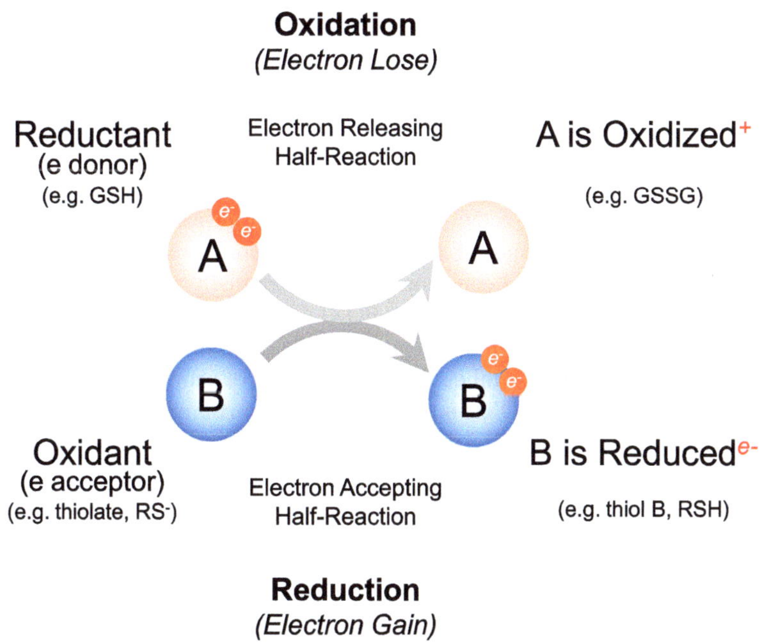

Fig. 4 Acidity and redox reactions. Redox reactions involve the transfer of electrons between species, where one species is oxidized (loses electrons) and another is reduced (gains electrons). The equilibrium stability of a redox reaction is subject to the redox potential of the couplers and the environment in which the reaction takes place. For redox reactions with protons as catalysts and/or as products, the presence of protons shifts the equilibrium of the redox reaction, thereby promoting the reduction half-reaction and/or maintaining the reduction state by providing protons and stabilizing reduced species. Thus, acidity favors the generation or retention of reductants on the right side of the equilibrium and is pro-reductive

molecules (Dereven'kov et al. 2013) and promote lipid peroxides initiation and propagation, whereas in alkaline conditions, lipid peroxyl radicals and the derivatives hydroperoxyl fatty acids and related free radicals tend to be hindered due to decreased lipid peroxidation rates (Ayala et al. 2014). Thus, maintaing a reducing state of a set-point of acidic balance while preventing the generaton of deleterious ROS in pH-sensitive redox coupling networks is essential for physiological processes in the epididymis. The full picture of potential regulatory mechanisms involved in pH-redox interplay in the epididymis also needs further exploration.

To maintain acid-base homeostasis, specific sets of genes are expressed in specific cells of a tissue. In the epididymal epithelium, acid-base homeostasis is maintained by specific isoforms of proton pump vacuolar ATPases (V-ATPase), which are abundantly expressed in the acid-secreting clear cells and moderately in the solute-transporting principal cells (Breton and Brown 2013; Breton et al. 1996; Shum et al. 2008, 2011b). Epididymal lumen is also acidified by bicarbonate cotransporters, including cytosolic and plasma membrane isoforms carbonate anhydrases, which work jointly to maintain low HCO_3^- concentration in the lumen (Pastor-Soler et al. 2005). Another set of gene expression contributing to acidification is the sodium-hydrogen exchangers (NHEs). NHE1 and NHE2 are expressed in the basal side and apical side, respectively, of epididymal epithelial cells (Chew et al. 2000). NHE3 also participates in the acidification of epididymis, via pathways in the efferent duct epithelial cells (Leung et al. 2001), involving the cAMP and adenosine signaling in the apical acid secretion of epididymal principal cells (Belardin et al. 2024). These proton pumps and pH-sensitive transporters expressed in the two major types of epididymal epithelial cells contribute differentially to acidify the luminal microenvironment (Park et al. 2017). The FOXI-1-knockout male mice are infertile due to impaired luminal acidification and V-ATPase expression (Blomqvist et al. 2006; Vidarsson et al. 2009). Although the mechanism of action of FOXI-1 remains unknown, its family analog FOXO is under the regualtion of redox stress (Klotz et al. 2015). In addition, genetic ablation intervention or pharmacological inhibition of proton secretion also led to male infertility (Liu et al. 2022; Mumtaz et al. 2022; Yeung et al. 2004). And in fact, these proton transporters are subjected to redox regulation, as reported in other biological systems (Akram et al. 2006; Khan et al. 2022; Miner et al. 2019; Seidel et al. 2012). As an example, the insertion of V-ATPase into plasma membrane for proton secretion is reciprocally regulated by actin cytoskeleton remodeling (Shum et al. 2011a), which is subjected to redox regulation such as NOXs (Wilson et al. 2016).

While the harmful effects of oxidative stress on epididymal sperm maturation have been known (O'Flaherty 2019; Vernet et al. 2004), medical antioxidant strategies can reduce sperm DNA fragmentation (Greco et al. 2005); however, this is not always beneficial, and the impact of hyper-reductive stress on epididymal sperm maturation remains unclear (Baskaran et al. 2020a; Henkel et al. 2019; Juarez-Rojas et al. 2021; Qazi et al. 2019; Sheweita et al. 2022; Vernet et al. 2004). The hyper-productive stress in the epididymal microenvironment is expected to be as harmful as oxidative stress and may impair sperm quality, especially as sperm DNA stability and fertilizing activities have been reported to be regulated by redox stress (Chen et al. 2012; Panner Selvam et al. 2020; Wang et al. 2009; Ward 2009). Therefore, it can be inferred that an imbalance in the pro-reductive microenvironment, whether it is hyper-reductive or excessive oxidative stress, that fails to meet the redox set point for functional and metabolic homeostasis of epididymal sperm maturation may lead to infertility, especially idiopathic male infertility, which may result in up to approximately 40% of pathological cases (Corsini et al. 2022).

The interplay of acidity and redox balance has been well studied in intracellular organelles (Bhattarai et al. 2021; Kellokumpu 2019; Lennicke and Cochemé 2021; Lennicke et al. 2015; Sies et al. 2024). Most of the redox couples are pH sensitive, and the equilibrium of their redox reactions is subjected to the regulation of acidity. This denotes that redox homeostasis is regulated through the protonation state (i.e., pKa) of redox pairs (Sies et al. 2024). H_2O_2, as an evolutionary molecule, also serves in the epididymis to protect spermatozoa from free radicals by rapidly dismutating ROS by SOD enzyme. H_2O_2 is then dissipated to water by various enzymes in different sub-cellular compartments, including catalase CAT, GPX system, and

PRDX system, as well as non-enzymatic bioactive substances such as GSH and vitamins (see Fig. 3). Chemically, the weakly acidic environment in epididymal lumen fluid can stabilize H_2O_2, and thus its signaling lifetime is expected to be longer compared with slightly alkaline plasma fluid. Another example is the thiol group in molecules, such as selenocysteine; the ratio of the protonated thiol form and deprotonated thiolate form can affect the properties and reactivities of molecules containing them. Thiols are primary targets of cellular redox modifications, mainly through thioredoxin reductases TRDs and glutathione peroxidases GPXs. The oxidized thiol group can be rapidly and efficiently reduced back to protonated form by the reducing downstream effectors of NADH- and NADPH-redox systems. These two processes regenerate the active reducing agents, thioredoxin and glutathione GSH, thereby establishing a highly effective defense line against ROS (Robaire and Hinton 2015; Sies et al. 2024; Xiao and Loscalzo 2020). Acidic luminal microenvironment in the epididymis favors the protonated GSH form and therefore forms a natural physiological defensing mechanism against various harmful hazards and can protect sperm.

3.3 The Decalcitonic Epididymal Luminal Microenvironment Is Interlinked with the Pro-reductive Acidic Conditions and Has Significant Implications for Male Reproductive Health

Redox and Ca^{2+} interplay. Redox stress is mutually associated with the mobilization of Ca^{2+} in cells as both are intracellular and extracellular signal messengers. Defects in the redox and Ca^{2+} interplay can lead to dysfunctions and pathological conditions (Gorlach et al. 2015; Hidalgo and Donoso 2008). The membrane Ca^{2+} permeable transporters localized in plasma membranes and membranes of intracellular organelles work in a concerted manner to maintain the tightly controlled Ca^{2+}-free concentrations ($[Ca^{2+}]$) in the specific extracellular and intracellular compartments. For instance, depending on cell types, the resting intracellular $[Ca^{2+}]$ is ~10^{-7} M, which is 10^4 times lower than $[Ca^{2+}]$ in extracellular milieu (~10^{-3} M), whereas the stored $[Ca^{2+}]$ in intracellular organelles can also reach between ~10^{-4} and ~ 10^{-3} M, depending on specific compartments. It is known that excessive levels of ROS cause oxidative stress, which in turn cause Ca^{2+} influx into the cytoplasm from the extracellular environment and from the endoplasmic reticulum or sarcoplasmic reticulum (ER/SR) through the cell membrane and the ER/SR channels, respectively. Rising $[Ca^{2+}]$ in the cytoplasm causes Ca^{2+} influx into other subcellular compartments, such as mitochondria and nuclei, and have a more long-term effect due to the influence of genetics in these organelles. The negative charges of ROS, especially free radicals, are highly reactive to mobilize any Ca^{2+} ions stored or trapped in organic matters (e.g., proteins or lipids or membranes) temporarily. This is good and fast enough to release Ca^{2+} from the stable status to mobile status, whereas

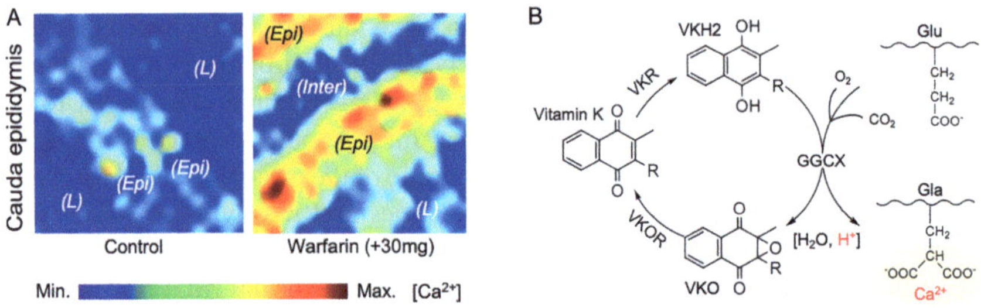

Fig. 5 Interplay of acidity and Ca^{2+} homeostasis in the epididymis revealed by carboxylation of MGP by the pH-dependent vitamin K2 redox-dependent activity of GGCX. (**a**) Synchrotron-based elemental X-ray fluorescence imaging of relative Ca^{2+} contents map in the cauda epididymis of rats from control group or vitamin K-cycle inhibitor warfarin-treated group. *(L):* lumen; (Epi): epithelium; (inter): interstitial space of epithelium. Control group: wild-type rats supplemented with vitamin K1; treatment group with 30 mg warfarin plus vitamin K1 supplement. (**b**) Both proton and Ca^{2+} ions are involved in the vitamin K redox-cycle dependent GGCX carboxylation function, in which protons serve as catalyst, whereas Ca^{2+} ions are the target to be chelated that in turn can affect the homeostasis of the redox-dependent GGCX-mediated carboxylation process and is subject to the influence by the pro-reductive condition in the epididymis. (Images are adapted from (Ma et al. 2019; Shum et al. 2022) with permission and modifications)

mobilized Ca^{2+} ions are indeed hazardous and dangerous, causing ROS-induced pathological conditions.

In the epididymal lumen, extracellular matrix [Ca^{2+}] decreases along the tubule of epididymis, from ~$1.3x10^{-3}$ M in the initial segment and to $0.25x10^{-3}$ M in the cauda region, which is obviously lower than that in the blood constantly with ~$2.5x10^{-3}$ M free Ca^{2+} in plasma (Hong et al. 1984; Levine and Marsh 1971; Weissgerber et al. 2011) (also see review in (Shum et al. 2022)). Previous work in our laboratory found that defects of the carboxylation of matric Gla protein (MGP) by gamma-glutamyl carboxylase (GGCX), a vitamin K redox cycle dependent enzyme, by inhibition with pharmacological tool warfarin leads to increased Ca^{2+} in the epididymal lumen (Fig. 5), accompanied increased Ca^{2+} in the epididymal fluids and epithelial cells (Ma et al. 2019). nA SNP mutation of GGCX is associated with the risk of asthenozoospermia in human (Ma et al. 2019), which is associated with the carboxylation of MGP and its chelation property of Ca^{2+} in the matrix of epididymal lumen (Shum et al. 2022). These findings suggest that the redox-dependent vitamin K2-dependent regulation of Ca^{2+} homeodynamics through carboxylation of MGP via GGCX plays a vital role in sperm maturation by maintaining low calcium luminal microenvironment and scavenging metabolites remnants in the epididymis.

The interplay of acidity and cellular Ca^{2+} homeostasis is known, and acidic intracellular organelles serve as Ca^{2+} stores, such as lysosomes (Lloyd-Evans and Waller-Evans 2020; Nunes and Demaurex 2014). However, how the extracellular acidity and Ca^{2+} contents influence intracellular acidity or vice versa remain largely unclear. In case of epididymal luminal microenvironment, our previous studies have shown that the interplay of luminal acidity and Ca^{2+} homeostasis in the rat epididymis involves vitamin D-dependent TRPV6-TMEM16A coupler and vitamin K2 (VK2)-dependent GGCX-MGP in a spatial expression complementary pattern (Gao et al. 2016; Ma et al. 2019; Shum et al. 2022). In the rat epididymal principal cells, the pH-sensitive epithelial Ca^{2+} channel TRPV6 participates in Ca^{2+} reabsorption in the rat epididymis and anionic secretion, which in turn drives the luminal fluid secretion of epididymis (Cheung et al. 2005; Wong et al. 2002). Thus, TRPV6 can electrically and reciprocally use Ca^{2+} to couple with the Ca^{2+}-dependent chloride channel TMEM16A in the principal cells in the very proximal caput and very distal cauda regions of epididymis. However, the TRPV6-TMEP16A coupler is absent in the length of regions of epididymis, where mostly Ca^{2+} is supposedly to be reabsorbed. This study led us to the subsequent studies and identified that the VK2 redox-cycle-dependent carboxylation of MGP by GGCX enzymatic activity is an additional pathway in the caput and corpus regions for Ca^{2+} homeodynamic regulation, where the TRPV6-TMEM16A coupled complex is absent (Ma et al. 2019). Intriguingly, the acidity functions as a catalytic factor of the pan-vitamin K redox cycle-dependent GGCX carboxylation process and thus the whole process can be enhanced by the acidity of the reaction environment, as illustrated in Fig. 5b. It is known that the pH sensitivity of ROS radicals towards complexes' chelating property with $Ca^{2+}$is an important consideration, as the pH can significantly affect the formation and stability of these complexes, such as GGCX and its final product carboxylated MGP, as demonstrated in our previous studies (Ma et al. 2019; Shum et al. 2022). It has been known that EVs are also annotated as matrix-calcifying vesicles (Lotvall et al. 2014), such as the EVs derived from vascular system, consistent with the fact that these EVs participate in extracellular calcium homeostatic regulation and calcification processes (Kapustin et al. 2011, 2015; Reynolds et al. 2004). In the epididymis, the presence of MGP-positive vesicle-like structures on the surface of epididymal spermatozoa and the occurrences of stress granules after the warfarin-induced interruption of VK2-dependent GGCX activity are in line with the involvement of epididymal EVs in the interplay of redox and acid-base balance in Ca^{2+} homeodynamics in the luminal microenvironment, as illustrated in Fig. 6. Moreover, it is known that MGP is synthesized locally to counterbalance ectopic mineral deposition (Murshed et al. 2004). The authors speculate that the carboxylation physiology of sperm maturation in the epididymal lumen, which may or may not involve Ca^{2+}-dependent epididymal EVs, may be independent of the carboxylation physiology in somatic systems. In this regard, relevant male fertility therapeutics, or as the other side of the coin, male contraceptive interventions, deserve dedicated exploration. Indeed, EVs are increasingly used for diagnostic and/or therapeutic applications in both reproductive and non-reproductive fields (Goss et al. 2023; Kumar et al. 2024a; Quesenberry et al. 2015; Skog et al. 2008).

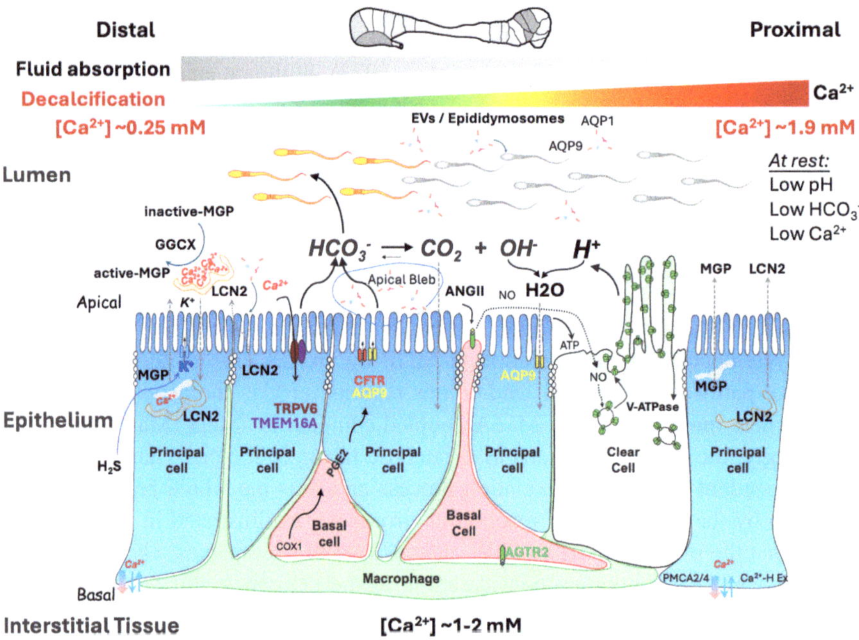

Fig. 6 Epididymal epithelial cells and extracellular carrier EVs in the epididymis: a glimpse of micro-environmental signaling interplay. The intriguing complex model includes at least several physiological processes to maintain redox balance in an acidic and decalcitonic microenvironment of epididymis: (1) anion secretion (e.g., HCO_3^-) via apically located CFTR (and/or other anion channels, such as TMEM16A) provides an ionic gradient that drives fluid transport through water channels, e.g., aquaporin9 (AQP9); (2) proton-pump V-ATPase acidifies the luminal content and promotes HCO_3^- hydrolysis for water and CO_2 absorption, an essential acidification process required for maintaining sperm in a quiescent state and to prevent premature activation. The acidification process is mediated constitutively either by the recycling of the V-ATPases that are abundantly expressed in clear cells or by the stimulation of luminal factors, such as ATP generated by principal cells or free radical nitric oxide (NO) from the basal cells upon stimulation by the angiotensin receptor type 2 (AGTR2); and (3) Ca^{2+} reabsorption either through the epithelial Ca^{2+} channel TRPV6 and Ca^{2+}-dependent Cl^-channel TMEM16A electrical coupler or via a process facilitated by GGCX-mediated carboxylation-dependent activation of MGP for Ca^{2+} chelation and simultaneous protein aggregation, such as lipocalins (e.g., LCN2); (4) redox-balance pathways involve the signaling molecule H_2S that stimulates potassium secretion by activating Ca-dependent and ATP-sensitive K-channels, and H_2O_2 can be transported across cellular membrane by water channels AQPs. More details and related references are discussed in the main contents. (Figure is adapted from Shum et al. 2022 with permission and modifications)

3.4 Epididymal EVs Adapt to Microenvironmental Signals

EVs, especially large EVs, are known to be released in response to stress (Jeppesen et al. 2023). Many, though not all, EVs can also be influenced by the pH and redox balance of the environment; thus, the redox and acid-base conditions can influence the biogenesis, release, and cargo loading of EVs (Yanez-Mo et al. 2015). The lipid bilayer of EVs and their membrane compositions like lipid rafts can undergo conformational changes at different pH levels, via their ionizable headgroups, influencing their stability and fusion capabilities. The pH of the extracellular environment can also influence the uptake of EVs by recipient cells and certain receptors may be more active or available at specific pH levels. Acidic environments can also trigger the release of cargo from EVs, such as proteins, RNA, and other biomolecules. As an example, it has been reported that environmental pH stress influences cellular secretion and uptake of EVs in in vitro cellular conditions (Nakase et al. 2021). The outer layer of EVs comprises membranous lipids, which can integrate with lipid rafts, sphingolipids, and specific proteins in particular types of EVs (Jeppesen et al. 2023; Mear et al. 2024). These integrated components are subjected to redox modifications. In fact, lipid rafts are known as redox-sensitive hubs and can trigger adaptive signaling and make some preconditioning conditions (Bhat et al. 2018; Jin et al. 2011). In the acidic, decalcitonic, and pro-reductive conditions of epididymis, EVs are therefore tailored to carry their loaded messages and to be formed and released for the communication between epididymal sperm and epithelial cells in the specialized luminal microenvironment. In other words, the EVs and microenvironmental signals (i.e., redox status, acid-base balance, and calcium gradients of the epididymal fluid) interact with each other to maintain the homeodynamics of luminal conditions. This intriguing interplay can mutually influence their roles and those of epithelial cells during sperm maturation and subsequent reproductive events.

Some adaptation characteristics of epididymal EVs to microenvironmental signals. In the epididymis, although there are still many unanswered questions about the physiological requirements and regulatory mechanisms of EVs' adaptation, the pioneering studies by Sullivan and colleagues on the response of EVs to the epididymal microenvironment have provided some important cues (Frenette et al. 2002). Their studies on the exchange of bovine epididymal EVs and sperm revealed that the protein exchange promoted by the interaction between the EVs and sperm is a joint effect of temperature and pH (Frenette et al. 2002). These researchers also found that addition of Zn^{2+} ions to the co-incubation medium can enhance the interaction of EVs and sperm. In the same study, Sullivan et al also found that transfer of EVs appears to be temperature sensitive, being more efficient at 32–37 °C than at 22 °C. The transfer of proteins to sperm is also pH-dependent, with the optimal pH for transfer being mild acidic, i.e., 6.0–6.5. They also investigated the effect of divalent cations with millimolar level of EDTA on the transfer of the proteins in cauda epididymal prostasome-like particles to caput epididymidal sperm and claimed that while Mg^{2+} and Ca^{2+} had no effect on the amount of protein that remained associated with spermatozoa after co-incubation, Zn^{2+} had a beneficial effect (Frenette et al. 2002). In the ram proteomics of epididymosomes, proton pump V-ATPase subunit A is found in a considerate level (Gatti et al. 2005). Consistent with pH adaptation, epididymosomes from clear cells, a major type of acid-secreting epithelial cells, in both mouse and human epididymis form a conserved component of epididymal EVs and specifically transfer the very long-chain fatty acid transporter SLC27A2 from clear cells to seminal spermatozoa (Barrachina et al. 2022). The isoform analogue of SLC27A2, SLC27A1, has been reported to be involved in redox sensing of oxidized VPS35 and impaired arginine transportation, leading to insufficient arginine for mitochondrial translation, which explains the decreased ROS levels in cancer cells (Zhang et al. 2025a). The transport of fatty acids across membranes is influenced by pH levels (Angelova et al. 2018), which is attributed to the influence onto the cargo loaded in the EVs. Given that clear cells express high levels of proton pump V-ATPase complex (including subunit A) and play an important role in proton-secretion and luminal acidification, this may imply that the membrane-associated V-ATPase complex may be involved in the reciprocal adaptive response of epididymal EVs to pH and other signals. In view of that V-ATPase in the yeast can be blocked by redox-active cation copper (Miner et al. 2019), the interplay of redox and pH in the epididymal EVs might be glimpsed in their interactions and adaptations.

Some pH-dependent redox sensitivity of EVs and their characteristics in the epididymis. In redox reactions, reduction and oxidation reactions occur and tend to reach an equilibrium depending on the gradients of reactants and products and the conditions. In fact, in the chemical world, redox reactions occur through electron transfer between reactant species and are influenced by proton gradients, i.e., acidity. On the other hand, although the overall influence of acidity in redox reactions shows a tendency to pro-antioxidation, redox reactions are also influenced by the net input of antioxidant-antireductant system. For EVs, their membrane lipids are redox sensors through their phospholipid headgroups (Angelova et al. 2018). When the lipid components on the membranes of EVs are exposed to ROS or other oxidizing agents, for example, sphingolipids in lipid rafts (Bhat et al. 2018) or PUFAs, like eicosanoids (Boilard 2018), lipid peroxidation can occur. The influence of pH on lipid peroxidation is indeed significant, with acidity and alkalinity playing opposing roles in the initiation, propagation, and termination of peroxidation reactions (Wei et al. 2022). Thus, in the protective pro-reductive acidic environmental conditions in the epididymis, it is expected that acidity can prevent ROS propagation, through acid-catalyzed rearrangement of peroxides and the protonated form of GSH of the key cellular GSH/GSSG redox couple. Although acidity reduces the overall antioxidant capacity of GSH redox cycle by reducing the GPX activity and GSH reactivity (Schafer and Buettner 2001), paradoxically, GR, glutathione reductase, functions optimally under slightly acidic conditions (~6.5–7.5) (Can et al. 2010), so by supporting GR activity, a good pool of active GSH can be maintained in the epididymal microenvironment to sustain the protective-redox buffering (Kaneko et al. 2002).

Given that redox regulatory couples utilize specialized EVs to participate in their unique intercellular communications and physiological functions (Barrachina et al. 2022; Nixon et al. 2018; Sullivan 2015), it has been reported that epididymosomes are heterogeneous and distinct types of epididymosomes express unique markers (Sullivan 2015) and can be derived from specific cell types (Barrachina et al. 2022). About one-third of the epididymosomal proteins in the mouse epididymosomes were enzymatic proteins, and most of them are associated with the redox couples, including GGTs, GPXs, GR, GSTs, TXNs, TRDs, and SOD (Nixon et al. 2018).

In some specialized EVs of non-reproductive system, NADPH oxidases have been reported, which can modulate the function of exosomes per se by generation of superoxide and subsequent downstream ROS (Gaspar et al. 2021; Janiszewski et al. 2004; Krishnamoorthy and Chang 2018). In the epididymis, caput sperm have been reported to generate some redox flux to themselves through the activity of NADPH oxidases (Aitken and Vernet 1998), and the functional roles of NOX family of NADPH oxidases in ROS generation in human spermatozoa have been reported (Cui et al. 2024; Musset et al. 2012), although the involvement of NOX oxidase in epididymal EVs has not been found. Nevertheless, these enzyme-containing epididymal EVs suggest that they are redox active and that the enzymes in the cargo of EVs play an important role combating stress in the epididymis and associated reproductive events (Gervasi and Visconti 2017).

Other types of specicalized pH-sensitive redox-associated EVs are also present in the epididymosomes, such as the H_2O_2-permeabble water channel aquaporins like AQP1 and AQP9, and glutathione downstream effectors such as cysteine- and glycine-rich protein 1 (CRSP1) and cysteine-rich secretory proteins (CRISP1) (Nixon et al. 2018). As aforementioned, H_2O_2 is a central redox-sensitive signaling molecule (Sies et al. 2024), together with NO and H_2S that are involved in synaptic vesicles for intercellular interactions (Chatterjee et al. 2020; Cirino et al. 2023), all of which are affected by acid-base balance. It is known that water channels aquaporins can facilitate the passage of water molecules (H_2O) across the plasma membrane, which was demonstrated by Agre's group in 1992 for which they were awarded the Nobel Prize in 2003 (Agre 2004). Strikingly, specific AQPs, called peroxiporins, can also transport H_2O_2 molecules across plasma membranes (Henzler and Steudle 2000), rendering the role of AQPs to play a functional role in H_2O_2 translocation (Bienert and Chaumont 2014) and therefore redox regulation (Sies et al. 2024). In the male reproductive tract, specific AQPs are expressed in specific cell types in different regions of epididymis and in the upstream organ effect ducts and the downstream organ vas deferens (Da Silva et al. 2006). For example, AQP9 is expressed in apical stereocilia principal cells but not clear cells in all regions of epididymis, whereas in the corpus and cauda epididymidis, AQP5 is also expressed in the apical stereocilia of principal cells, while AQP3 as well as AQP8 are expressed in the basal cells (Da Silva et al. 2006). Given that AQPs can transport H_2O_2 (Bienert and Chaumont 2014) and AQPs can be loaded in EVs (Clarke-Bland et al. 2022; Oshikawa et al. 2016; Oshikawa-Hori et al. 2019), although whether any AQP is present in epididymal EVs remains to be confirmed, EVs carrying these redox-active components are expected to contribute to the redox homeostatic regulation of the epididymal-dependent reproductive processes, including sperm capacitation and acrosome reactions during fertilization of oocyte (Ikawa et al. 2010).

4 Nutrient Balance and Involvement of EVs in Cellular Interplay During Epididymal Sperm Maturation

In the epididymal lumen, spermatozoa are nurtured in a specialized nutritional milieu, which provides essential raw materials and energy required for their maturation and the physiological functions of both somatic and gametic cells. Nutrients are generally classified into two main categories. One category is macronutrients that are required in large amounts and are primarily used for energy and structural functions, including carbohydrates, proteins, lipids and water; the other category is the micronutrients that are required in smaller amounts but critical for cellular processes, including organic vitamins and inorganic minerals. Malnutrition is detrimental to health, while insufficient nutritional supplies, whether macronutrients or micronutrients, leads to unhealthy conditions, and excessive nutrition is also harmful. For example, taking excessive amounts of vitamins can cause hazardous consequences, and inappropriate supplementary of vitamins can be harmful and may associate with a higher mortality risk (Loftfield et al. 2024). Therefore, a well-balanced nutrition is the key. In this section, although not intended to be comprehensive, the authors focus on micronutrients and their relevance to EVs in the epididymis; regarding macronutrients in the epididymis, refer to other comprehensive review articles (Gervasi and Visconti 2017; Hermo and Robaire 2002; Myers and Cornwall 2024; Pastor-Soler et al. 2005; Robaire and Hinton 2015; Shum et al. 2022; Turner 2002; Wong et al. 2002; Zhou et al. 2018).

4.1 Special Nutrition Is Required in the Epididymal Luminal Microenvironment

Providing balanced nutrients and supporting necessary redox metabolism are the fundamental roles of the epididymal epithelial cells in promoting sperm maturation. Even when nurtured under normal physiological conditions, maintaining nutritional balance is one of the major stress challenges faced by sperm during epididymal maturation. Given that low levels of redox input are required to promote sperm maturation (Aitken and Curry 2011; Aitken and Vernet 1998; Henkel et al. 2019; Moon et al. 2024) and that imbalanced redox can be hazardous, it is necessary to continuously monitor the redox state between sperm and epithelial cells (Aitken and Drevet 2020; Sies et al. 2024). In addition to the extensive exchanges of nutrients and raw materials between sperm cells bathed in the lumen and the nurturing epididymal epithelial cells, the luminal matrix plasma not only provides physiological supports but also serves as a provider of macro- and micronutrients with desirable dosages (Robaire and Hinton 2015; Shum et al. 2022; Weigel Munoz et al. 2024). For example, compared to blood plasma, epididymal luminal fluid has a higher potassium-to-sodium ratio, lower luminal pH, and lower calcium level (Breton et al. 1996; Jenkins et al. 1980; Levine and Marsh 1971; Robaire and Hinton

2015; Shum et al. 2022; Turner 2002), which maintain spermatozoa in a dormant state before activation (Hamamah and Gatti 1998; Shum et al. 2009). Thus, a specialized nutritional milieu is established in the epididymal luminal microenvironment to meet the physiological needs.

Two basic nutritional needs are required in the epididymis. For sperm to mature in the epididymis, the first principle for them is to survive well, then to mature with necessary functional modifications (Chen et al. 2016a, 2021; Chu et al. 2019; Conine et al. 2018; Gervasi and Visconti 2017; Hermo and Robaire 2002; Hinton 1990; Myers and Cornwall 2024; Pastor-Soler et al. 2005; Robaire and Hinton 2015; Sharma et al. 2016, 2018; Shum et al. 2022; Tomar et al. 2024; Turner 2002; Wang et al. 2023; Wong et al. 2002; Zhang et al. 2019, 2021b). Basically, during the process of sperm maturation, the biological needs from sperm can be categorized into two major activity of requirements: one is to rely on continuous metabolic energy to survive, and the other is to mature to acquire competent fertilization ability. For survival mode, they need a sustainable supply of redox balanced nutrients and inputs; for maturation mode, they learn external environmental information from the luminal milieu by exchanging with epithelial cells reciprocally. In the epididymis, sperm cells obtain the nutrients from the luminal fluids, including macronutrients such as carbohydrates, proteins, and lipids; the micronutrients including vitamins and essential inorganic elements; and some small molecules like amino acids and small RNAs that may vary in a wide range of quantities. Epididymal somatic cells obtain their nutrients the circulation systems and/or the luminal fluids present in the reproductive tract.

4.2 *Essential Micronutrients for Spermatozoa in the Epididymal Microenvironment*

Vitamins are vital organic micronutrients that cannot be synthesized endogenously or in sufficient quantities and therefore need to be obtained from the diet. There is now a consensus on the definition of a vitamin, which must be an organic compound that cannot be synthesized in the body and must be obtained in the diet, and its presence in the diet is essential in preventing disease. In mammals, at least 21 vitamins and their direct derivatives and vitamin-like essential nutrients are required. Under physiological conditions, almost all these essential vitamins, except vitamin B7, are reversibly active in redox cycling reactions and are commonly considered antioxidants due to their free radical scavenging properties (Table 1 and references wherein). As for vitamin B7, also known as biotin or vitamin H, its conversion to ester carboxylate anion form is irreversible by reduction reaction (Lauw et al. 2013). Most of the redox-active vitamins act as antioxidants when they neutralize free radicals by donating their own electrons, whereas some of them can be anti-reductants by acting as electron acceptors to prevent progressive adverse consequences of free radicals. Under certain extreme conditions, some antioxidant

Table 1 Redox activity, pH-sensitivity, and calcium chelation property of key forms of vitamins in mammalian systems

Vitamins	Solubility	Redox-ensitive[a]	Redox reaction nature	Antioxidant capacity[b]	Reactant or co-factor involved[b]	Side effect at odd-dose[b]	pH-sensitivity[b]	Ca-chelation[b]	Metal Dependency	References
Vitamin A										
Retinol	Fat soluble	Yes/active	Antioxidant defense (ROS scavenging)	$^1O_2$ quencher	Metal-binding enzymes/ proteins	Pro-oxidant & toxic at high intakes	Moderate; stable at pH 6–8	Yes (direct)	Indirect (Zn, Cu)	Martinez et al. (2008)
Retinal	Fat soluble	Yes/active	Antioxidant defense (ROS scavenging)	$^1O_2$ quencher	Metal-binding enzymes/ proteins	Pro-oxidant & toxic at high intakes	Moderate; stable at pH 6–8	Yes (direct)	Indirect	Martinez et al. (2008)
Retinoic acid	Fat soluble	Yes/active	Antioxidant defense (ROS scavenging)	$^1O_2$ quencher	Metal-binding enzymes/ proteins	Pro-oxidant & toxic at high intakes	Moderate; stable at pH 4–10; pKa~4.8	Yes (direct)	Direct (Zn, Fe, Cu)	Martinez et al. (2008)
Pro A (carotenoids)	Fat soluble	Yes/active	Antioxidant defense (ROS scavenging)	$^1O_2$ quencher	Metal-binding enzymes/ proteins	Pro-oxidant & toxic at high intakes	Low; stable at wide pH ranges	Yes (direct)	Indirect (Zn, Fe, Cu)	Martinez et al. (2008), Asensi-Fabado and Munne-Bosch (2010)
Vitamin B										
B1 (thiamine)	Water soluble	Yes/active	Pro-oxidant (catabolism; cofactor)	Scavenger of $O_2^{\bullet-}$, $^{\bullet}OH$	Thiamine-dependent enzymes	Readily excreted in urine & generally safe	Sensitive; stable at pH 5–7	No (indirect)	Indirect	Asensi-Fabado and Munne-Bosch (2010), Huang et al. (2010), Mrowicka et al. (2023), Hu et al. (1995) and Hrubsa et al. (2022)

B2 (riboflavin)	Water soluble	Yes/ unknown	Antioxidant defense (redox mediator, respiration precursor)	Scavenger of ROS	Riboflavin-dependent enzymes (e.g., cP-450, Xanthine oxidase)	Readily excreted in urine & generally safe	Moderate; stable at pH 4–8	No (indirect)	Indirect (Mg, Zn, Fe, Cu)	Asensi-Fabado and Munne-Bosch (2010), Hrubsa et al. (2022)
B3 (niacin)	Water soluble	Yes/ unknown	Pro-antioxidant (redox cycle; electron transfer, precursor)	Scavenger of ROS	Metal-dependent enzymes	Hypolipidemic in gram doses	Low; stable at wide pH ranges	No (salt form)	Indirect (Mg, Zn, Fe, Cu, Mn)	Asensi-Fabado and Munne-Bosch (2010), Hrubsa et al. (2022)
B5 (pantothenic acid)	Water soluble	Yes/ unknown	Pro-oxidant (acyl-group transfer)	CoA precursor	Aa a component of all CoA-dependent proteins	Readily excreted in urine & generally safe	Sensitive; stable at pH 5–7	Yes (direct, favored at high pH)	Direct (Ca); Indirect (Mg, Mn)	Asensi-Fabado and Munne-Bosch (2010), Hu et al. (1995), Hrubsa et al. (2022) and Corinti et al. (2021)
B6 (pyridoxine)	Water soluble	Yes/low	Pro-oxidant (cofactor)	Scavenger of $O_2^{\bullet-}$, $^{\bullet}OH$	Various metabolic enzymes	Readily excreted in urine & generally safe	Moderate; stable at pH 5–8	No (indirect)	Unknown (Fe)	Asensi-Fabado and Munne-Bosch (2010), Hu et al. (1995), Franco et al. (2022) and Parra et al. (2018)
B7 (Biotin)	Water soluble	No/inert	Antioxidant defense (CO_2 group transfer, carboxylation)	None described	Two Mg-binding carboxylases and coenzymes	Readily excreted in urine & generally safe	Low; stable at pH 3–10	No (indirect)	Indirect (Mg, Mn)	Asensi-Fabado and Munne-Bosch (2010), Waldrop et al. (2012) and Lauw et al. (2013)

(continued)

Table 1 (continued)

Vitamins	Solubility	Redox-ensitive[a]	Redox reaction nature	Antioxidant capacity[b]	Reactant or co-factor involved[b]	Side effect at odd-dose[b]	pH-sensitivity[b]	Ca-chelation[b]	Metal Dependency	References
B9 (Folic acid)	Water soluble	Yes/ moderate	Antioxidant-favored & pro-oxidant (1C transfer)	Scavenger of $O_2^{\bullet-}$,$^{\bullet}OH$	Various metabolic enzymes; active form THF	Readily excreted in urine & generally safe	Sensitive; stable at pH 5–7	No (salt form)	Direct (Mg, Zn, Fe, Cu, Mn)	Asensi-Fabado and Munne-Bosch (2010), Franco et al. (2022) and Siatka et al. (2025)
B12 (Cobalamin)	Water soluble	Yes/ moderate	Antioxidant-favored & pro-oxidant (1C- transfer, catabolic radical-based)	Scavenger of $O_2^{\bullet-}$,$^{\bullet}OH$	B12-dependent enzymes (e.g., methionine synthase; vitamin B9)	Readily excreted in urine & generally safe	Low; stable at pH 3–10	Yes (direct; Co>Ca)	Direct (Co)	Randaccio et al. (2010), Dereven'kov et al. (2013)
Vitamin C										
Ascorbic acid	Water soluble	Yes/active	Antioxidant (pH-sensitive)	Scavenger of $^1O_2$, $^{\bullet}OH$ or $O_2^{\bullet-}$, oxidized Fe/Cu ions	Substrate of hydroxylase, oxygenase, and NADH- & GSH-dependent enzymes	Pro-oxidant with specific metal ions	Sensitive; stable at pH 5–7	No (salt form)	Direct & indirect (Zn, Fe, Cu)	Martinez et al. (2008), Hondal (2023) and Alves et al. (2020)
Vitamin D										
D2 (Ergocalciferol)	Fat -soluble	Yes/low	Antioxidant & pro-oxidant equilibrium (co-factor)	$O_2^{\bullet-}$ quencher	Metal-binding enzymes (e.g., SOD, catalase)	Pro-oxidant & toxic at high intakes	Low; stable at pH 3–10	Yes (direct)	Indirect (Fe, Cu)	Zmijewski (2022), Kidroni et al. (1983) and Wiseman (1993)
D3 (Cholecalcifeerol)	Fat soluble	Yes/low	Antioxidant & pro-oxidant equilibrium (co-factor)	$O_2^{\bullet-}$ quencher	Metal-binding enzymes (e.g., SOD, catalase)	Pro-oxidant & toxic at high intakes	Low; stable at pH 3–10	Yes (direct)	Indirect (Fe, cu)	Zmijewski (2022), Kidroni et al. (1983) and Wiseman (1993)

1,25$(OH)_2$ D3 (Calcitriol)	Fat soluble	Yes/low	Antioxidant & pro-oxidant equilibrium (co-factor)	$O_2^{\bullet-}$ quencher	Enzymes (e.g., SOD, catalase)	Pro-oxidant & toxic at high intakes	Low; stable at pH 3–10	Yes (direct)	Indirect (Fe, Cu)	Zmijewski (2022), Kidroni et al. (1983) and Wiseman (1993)
Pro D3 (7-DHC)	Fat soluble	Yes/ conditional	Reductant (UV-light dependent; electron-donor)	Absorb UV-B light on skin	Precursor of D3 and cholesterol	Pro-oxidant & toxic at high intakes	Low; stable at pH 3–10	No		Zmijewski (2022), Kidroni et al. (1983), Wiseman (1993) and Liu et al. (2011)
Vitamin E										
Tocopherols (α, β, γ, δ isoforms)	Fat soluble	Yes/active	Antioxidant (ROS anion scavenger)	ROS scavenging	Metal-binding organic matters	Toxic at high intakes	Low; stable at pH 3–10	No (indirect)	Indirect (Fe, Cu, Se)	Martinez et al. (2008), Hu et al. (1995) and Hondal (2023)
Tocotrienols (α, β, γ, δ isoforms)	Fat soluble	Yes/active	Antioxidant (ROS anion scavenger)	ROS scavenging	Metal-binding organic matters	Toxic at high intakes	Low; stable at pH 3–10	Yes (direct)	Indirect (Fe, Cu, Se)	Martinez et al. (2008), Hu et al. (1995) and Hondal (2023)
Vitamin K										
K1 (Phylloquinone, natural form)	Fat soluble	Yes/active	Antiradicals (redox, proton-coupled electron carrier)	Radical & peroxide quencher; GSH production; pro-oxidant at high levels	VK-dependent enzymes (GGCX, VKDR, some metal-binding enzymes)	Some risk potential, but harmful effect at extremely high intake	Low; stable at pH 3–10	Yes (direct)	Indirect (Fe, Cu, other metals)	Rishavy and Berkner (2012), Ivanova et al. (2018)

(continued)

Table 1 (continued)

Vitamins	Solubility	Redox-ensitive[a]	Redox reaction nature	Antioxidant capacity[b]	Reactant or co-factor involved[b]	Side effect at odd-dose[b]	pH-sensitivity[b]	Ca-chelation[b]	Metal Dependency	References
K2 (Menaquinones, natural forms)	Fat soluble	Yes/active	Antiradicals (redox, proton-coupled electron carrier)	Radical & peroxide quencher; GSH production; pro-oxidant at high levels	VK-dependent enzymes (GGCX, VKDR, some metal-binding enzymes)	Some risk potential, but harmful effect at extremely high intake	Low; stable at pH 3–10	No (indirect)	Indirect (Fe, Cu, other metals)	Rishavy and Berkner (2012), Ivanova et al. (2018)
Synthetic K3 (Menadione, synthetic form)	Water soluble	Yes/active	Antiradicals (redox, proton-coupled electron carrier)	Radical & peroxide quencher; GSH production; pro-oxidant at high levels	VK-dependent enzymes (GGCX, VKDR, some metal-binding enzymes)	Some risk potential, but harmful effect at extremely high intake	Redox activity at pH 7–8; dissipate at low pH	Yes (direct)	Direct (Fe, Cu, Zn, Ca, Mg, Cd); metal-dependent redox	Ivanova et al. (2018)
Vitamin-Like										
EGT (Ergothioneine)	Water soluble	Yes/active	Antioxidant	Radical & peroxide quencher; maintain labile Fe pool		Considered as safe at high doses	Stable in extreme pH		Direct (Co, Fe, Cu, Zn, Ni)	Hondal (2023), Borodina et al. (2020)

$^1O_2$: singlet oxygen radical; $O_2^{\bullet-}$: superoxide anion radical; $^{\bullet}OH$: hydroxyl radical

[a]Redox sensitive categorized as active, moderate, inert, or unknown based on the TEAC values reported or on redox values (Asensi-Fabado and Munne-Bosch 2010; Hu et al. 1995; Jones 2008)

[b]Unless otherwise annotated, general information generated by OpenAI via Poe

Table 2 Redox reactivity, pH-sensitivity, calcium chelation property, and metal dependency of key forms of non-vitamins in mammalian systems

Non-vitamins	Solubility	Redox-sensitive[a]	Redox reaction nature	Antioxidant capacity[b]	Reactant or co-factor involved[b]	Side effect at odd-dose[b]	pH-sensitivity[b]	Ca-chelation[b]	Metal dependency	References
Free radicals										
$^1O_2$; $O_2^{\bullet -}$; •NO; •NO_2; $^{\bullet}OH$; •HOO; •CO_3	Water soluble	Yes/active	Pro-oxidant (oxidative chain reactions)	No. oxidizing agents	Many oxidative organic and inorganic matters	Major source of harmful oxidative stress	High (vary)	Yes (direct)	Indirect (Zn, Cu)	Jones (2008), Lobo et al. (2010)
•CCl_3	Water soluble	Yes/active	Pro-oxidant (oxidative chain reactions)	No. oxidizing agents	Many oxidative organic and inorganic matters	Major source of harmful oxidative stress	Low (varies at low vs high pH)	Yes (direct)	Indirect	Jones (2008), Unsal et al. (2021)
Nicotinamide-conjugates (non-radical)										
NAD	Water soluble	Yes/active	Redox carrier	Maintain redox homeostasis	Coenzyme of redox metabolism, e.g., TCA cycle	Endogenous proteins & generally safe	High; stable at pH 2–6			Xie et al. (2020)
NADP	Water soluble	Yes/active	Redox carrier	Maintain redox homeostasis	Coenzyme of redox metabolism, e.g., reductive lipid biosynthesis	Endogenous proteins & generally safe	High; stable at pH 8; unstable in phosphate buffer			Xie et al. (2020)
Peroxides (non-radical)										
H_2O_2	Water soluble	Yes/active	Pro-oxidant (potent oxidant)	No. highly reactive oxidizing agent	Many oxidative organic and inorganic matters	Major source of harmful oxidative stress	Low pH: stronger pro-oxidant, less labile; high pH: dissipate	Yes (direct)	Direct (Fe, Cu)	Jones (2008), Lobo et al. (2010) and Keller et al. (2015)

(continued)

Table 2 (continued)

Non-vitamins	Solubility	Redox-sensitive[a]	Redox reaction nature	Antioxidant capacity[b]	Reactant or co-factor involved[b]	Side effect at odd-dose[b]	pH-sensitivity[b]	Ca-chelation[b]	Metal dependency	References
Lipid peroxides	Fat soluble	Yes/active	Pro-oxidant (potent oxidant)	No. Highly reactive oxidizing agent	Lipophilic organic matters (e.g., PUFAs)	Major source of harmful oxidative stress	Low pH: more labile; high pH: dissipate	Yes (direct)	Direct (Fe, Cu)	Jones (2008), Lobo et al. (2010) andKeller et al. (2015)
Peroxynitrite (ONOO-)	Water soluble	Yes/active	Pro-oxidant (potent oxidant)	No. Highly reactive oxidizing agent	Many oxidative organic and inorganic matters	Endogenous; major source of harmful oxidative stress	Low pH: more labile; high pH: dissipate	Yes (direct)		Jones (2008), Lobo et al. (2010) and Keller et al. (2015)
Hydroperoxy fatty acids (non-radical)										
PUFA-derived hydroperoxides	Water soluble	Yes/active	Pro-oxidant (potent oxidant; ferroptosis inhibition)	Free radical trapping	GPX and GSH, membranes, Fenten reactants, eicosanoids	Harmful oxidative stress	Low pH: more labile; high pH: dissipate	Yes (direct)		Jones (2008), Lobo et al. (2010) and Keller et al. (2015)
Lipid-derived aldehydes	Various	Yes/ conditional	Reversible redox cycle (reactive carbonyl group)	Scavenging ROS (or pro-oxidant)	4-HNE-deriving PUFAs	A source of oxidative stress	Low pH: more labile; high pH: dissipate	Yes (direct)		Jones (2008), Keller et al. (2015)
Aromatic derivatives (non-radical)										
Quinones	Various	Yes/ conditional	Reversible redox cycle (proton-coupled electron transfer)	Antioxidants (or pro-oxidant); radical scavenging; preventing peroxidation	Coenzyme Q: mitochondrial inner membrane; quinone reductase/ oxidase; GGCX	Toxic at non-physiological high range	Low—moderate; various	Yes (direct)		Jones (2008)

Thiol-systems (non-radicals)										
Disulfides	Various	Yes/active	Redox (reversible disulfide bonds)	Free radical scavenging; reducing agent; regenerating redox status	Numerous thiol-group containing proteins	Liver and kidney toxicity at high dose	High (pH-dependent redox equilibrium)	Yes (direct)		Jones (2008), Keller et al. (2015)
H_2S	Both	Yes/active	Antioxidant (reducing agent)	Maintaining GSH pool; post-translational reduction modifications	Many organic matters	Toxic at non-physiological high range	Significant (pH-dependent redox equilibrium)	Yes (direct)		Jones (2008), Keller et al. (2015)
Glutathione (GSH)	Water soluble	Yes/active	Antioxidant (reversible redox)	Scavenging free radicals; constituting the GSH-GSSG pool	e.g., glutathione reductase	Generally safe but harmful at excessive supplement	Significant (favoring GSSG at low pH)	Yes (direct)		Jones (2008), Keller et al. (2015)
Thoredoxins (TRx)	Water soluble	Yes/active	Antioxidant (reversible redox)	Direct and indirect scavenging free radicals	e.g., thioredoxin reductase; calmodulin	Endogenous proteins & generally safe	Moderate (favoring oxidized TRx at low pH)	Yes (direct)		Jones (2008), Keller et al. (2015)
Cysteine (Cys)	Water soluble	Yes/active	Antioxidant (reversible redox)	Scavenging free radicals; GSH synthesis	e.g., metal-sulfur cluster moieties of proteins or macromolecules	Endogenous proteins & generally safe	Significant (thiol at low pH; thiolate at high pH)	Yes (direct)	Metal chelation	Jones (2008), Keller et al. (2015)
Redox-active transition metals and elements (non-radical)										
Copper (Cu^+ and Cu^{2+})	Water soluble	Yes/active	Redox (reversible electron transfer)	Cu(I): neutralize free radicals; ROS scavenger	Cu-containing enzymes (e.g., SOD)	Cu(II) can be toxic in excessive or imbalance	High (Cu (I) favors low pH, Cu(II) at high pH)	No (indirect)		Jones (2008), Tvrda et al. (2015)
Iron (Fe^{2+} and Fe^{3+})	Water soluble	Yes/active	Redox (reversible electron transfer; Fenton reaction)	Fe (II): neutralize free radicals; ROS scavenger	Ferritin, lactoferrin, transferrin, etc.	Toxic at non-physiological high range	High Fe (II) stable low pH, Fe (III) at high pH)	No (indirect)		Jones (2008), Tvrda et al. (2015)

(continued)

Table 2 (continued)

Non-vitamins	Solubility	Redox-sensitive[a]	Redox reaction nature	Antioxidant capacity[b]	Reactant or co-factor involved[b]	Side effect at odd-dose[b]	pH-sensitivity[b]	Ca-chelation[b]	Metal dependency	References
Other essential elements (e.g., Co, Mn, Se)	Water soluble	Yes/active	Redox (reversible electron transfer)	Co (II) or Mn (II): neutralize free radicals, activating enzymes	Co: e.g., vitamin B12, SOD; Mn: e.g., MnSOD; Se: e.g., GPXs	Harmful or toxic at non-physiological range	High (vary)	No (indirect)		Jones (2008), Ingold et al. (2018)
Other non-essential elements (e.g., As)	Water soluble	Yes/active	Pro-oxidant (electron transfer; potent oxidant)	None for As: an oxidizing agent binding to proteins (e.g., thiol groups)	e.g., pollutants	Poisonous even at small amount	High (pH-redox state-speciation)	Yes (direct)		Jones (2008), Kumar et al. (2024b)

[a]Redox sensitive categorized as active, moderate, inert, or unknown based on the TEAC values reported or on redox values (Asensi-Fabado and Munne-Bosch 2010; Hu et al. 1995; Jones 2008; Tvrda et al. 2015)

[b]Unless otherwise annotated, general information generated by OpenAI via Poe

vitamins can become pro-oxidants by shifting their electrochemical potentials of donating electrons to have the tendency to accept electrons, such as some water-soluble vitamins (Hu et al. 1995) and fat-soluble vitamins (Khatami et al. 2016). In the scenario of free radicals interactions, the term "antiradical" may be a better description for a redox-active substance capable of dual actions, which can either donate or accept electrons, as it can undergo either oxidation or reduction within the molecule, depending on specific biochemical conditions (Martinez et al. 2008). In view of the acidic and decalcified nature of epididymal microenvironment, the biophysical characteristics of vitamins toward pH sensitivity and calcium chelation are also summarized. To provide a more complete understanding of redox metabolism in mammalian systems, the redox reactivity, the pH-sensitivity, and calcium chelation properties of the major non-vitamins are also summarized in Table 2. In general, the functional roles of vitamins can be categorized into two groups based on their solubility: water-soluble vitamins and fat-soluble vitamins (Asensi-Fabado and Munne-Bosch 2010). Mineral micronutrients can also be categorized as redox-sensitive and redox-insensitive (Zoroddu et al. 2019). Their relevant functions on epididymal sperm and microenvironmental physiology and EVs are briefly discussed below.

4.3 Water-Insoluble Vitamins

(i) *Vitamin A*

Vitamin A refers to a group of fat-soluble retinoids that must be obtained entirely through diet—either as provitamin A carotenoids from plants or retinyl ester from animal tissues. These dietary forms are then converted into the performed vitamin A, mainly in the form of retinol, through a series of metabolic steps in the epithelium of intestine in humans (Orgebin-Crist et al. 2002). Due to their hydrophobic nature, retinoids require molecular chaperones—retinoid binding proteins (RBPs)—to ensure their proper storage, transport, and uptake in tissues. Excessive intake of vitamin A can be harmful and may lead to toxicity (Johansson and Melhus 2001; Schnorr et al. 2011). In the epididymis, a high level of secreted protein, epididymal retinoid acid-binding protein (ERABP), also known as a lipocalin, is present in the luminal fluid and plays important role in sperm maturation (Ong et al. 2000). A recent study reveals that circulating vitamin A retinol has causal effects in association with clinical phenotypes through the genetics of its binding proteins (Reay et al. 2024). Retinoids and RBPs have been reported to have functional roles in regulating EVs, contributing to the biology of stem cells, diabetic retinopathy, and inflammation (Fang and Liang 2021; Gr et al. 2025; Lin et al. 2021; You et al. 2021). In the epididymal EVs, retinoid-related enzymes have been detected, including retinol dehydrogenase ADH7 and retinal dehydrogenase ALDH1 (Gatti et al. 2005; Girouard et al. 2011; Keerthikumar et al. 2016; Skerget et al. 2015; Thimon et al.

2008). This highlights the potential role of retinoids via EVs in epididymal function and reproductive health.

(ii) *Vitamin D*

Vitamin D is another group of fat-soluble vitamins and functions as antioxidants in lipids and membranes, involving inhibition of iron-dependent lipid peroxidation, a hallmark of ferroptosis (Holick 2009; Wiseman 1993; Zmijewski 2022). The active form of vitamin D 1,25-dihydroxyvitamin D ($1,25(OH)_2D$), or known as calcitriol hormone, has a half-life on hourly basis, which is much shorter than the half-life of other inert forms of vitamin D, including D_2, ergocalciferol, and the D_3 cholecalciferol. Both vitamins D_2 and D_3 can be absorbed from diet, but they require two successive hydroxylations first in the liver on carbon 25 and then in the kidney on carbon 1 to form the provitamin D_3 (Pro D_3), which can also be converted from 7-dehydocholesterol (7-DHC) on the skin by sunlight. Pro D_3 is then converted to active calcitriol. In cauda epididymidal tissues, a vitamin D metabolite 24,25-dihydroxy form is found in high concentrations (Kidroni et al. 1983). 24,25-dihydroxy vitamin D is essential for bone formation (Ornoy et al. 1978) and acts as a marker of vitamin D catabolism, and its decreased concentration indicates chronic kidney dysfunction (Bosworth et al. 2012); however, its role in epididymal sperm maturation is still unclear. Vitamin D and cholesterol levels are correlated and vitamin D supplementation could help boost HDL in healthy individuals but reduce HDL and triglycerides in individuals with type 2 diabetes (Lu et al. 2024; Radkhah et al. 2023). While vitamin D insufficiency and deficiency is a prevalent pandemic healthy issue (Lupu et al. 2024), excessive intake of vitamin D through over-fortification of food can cause unhealthy hypercalciuria (Holick 2009). The role of vitamin D in calcium homeostasis and sperm function and male reproduction is well known (Blomberg Jensen et al. 2011; Blomberg Jensen et al. 2010; Jensen 2014; Yahyavi et al. 2024). In addition, the involvement of vitamin D and its receptor in the function of EVs and their contents and compositions have also been reported in non-reproductive somatic cells (Bisserier et al. 2022; Conte et al. 2021; Negri et al. 2022; Payet et al. 2025). In the EVs of non-reproductive cells or epididymal EVs, a low level of vitamin D binding protein (DBP) has been detected (Gatti et al. 2005; Girouard et al. 2011; Keerthikumar et al. 2016; Skerget et al. 2015; Thimon et al. 2008). Vitamin D and all its metabolites are bound to the chaperone DBP, which is also a member of the albuminoid family involved in the transport of fatty acids or hormones (Bouillon et al. 2019). This is also consistent with the fact that vitamin D and its binding protein DBP play an important role in epididymal function and reproductive health, either with or without EVs.

(iii) *Vitamin E*

Vitamin E has long been known as powerful fat-soluble antioxidants. This group of vitamins are categorized into tocopherols and tocotrienols, which can break hazardous radical chain reactions in the membranes and lipophilic matters (Hu et al. 1995; Martinez et al. 2008) and protect from ferroptosis (Hu et al. 2021). In addition to antioxidant and anticancer effects, the benefits of vitamin E in male reproductive health have also been documented (Jervis and Robaire 2004; Keskes-Ammar et al. 2003; Matorras et al. 2020; Mohammadi et al. 2022; Momeni and Eskandari 2012). In the epididymis, the measured levels of vitamin E were found to decline during epididymal transit, being significantly lower in spermatozoa from the cauda epididymidis, compared with cells from the caput region (Tramer et al. 1998). Excessive intake of water-insoluble vitamins, such as vitamin E, can lead to health complications, including the induction of reductive stress (Khatami et al. 2016; Perez-Torres et al. 2017), which may negatively affect male reproductive health (Henkel et al. 2019). For example, other redox-dependent substances—such as vitamin K precursors—need enzymatic conversion into their active functional forms. It is thought that excessive levels of vitamin E may interfere with these precursors and their associated enzymes, thereby impairing the coagulation process and increasing the risk of bleeding (Owen and Dewald 2025). Vitamin E treatment enhanced the function of dendritic cells via inhibition of SHP1 and increased antigen presentation by dendritic cells and their derived EVs to elicit anti-tumor immunity (Yuan et al. 2022); although thus far, no vitamin E in EVs has been reported. However, the lipophilic nature of vitamin E enables it to interact with EVs, thus preventing free radical chain reactions from harmful consequences, especially the protection of polyunsaturated lipids on the EVs. It is known that the lipid membrane composition of both epididymosomes and spermatozoa undergoes remodeling during epididymal maturation, and the proportion of lipid profiles is reversely correlated (Girouard et al. 2011; Rejraji et al. 2006). Compared with epididymosomes, the sperm membranes become more fluid, suggesting more polyunsaturated lipids (Rejraji et al. 2006). Consistently, the use of vitamin E in semen cryoprotection and sperm integrity has been documented (Kalthur et al. 2011).

(iv) *Vitamin K*

Vitamin K is a group of fat-soluble vitamins, which exists in two main natural forms, vitamin K1 (also known as phylloquinone) and vitamin K2 (which has multiple subtypes of menaquinones, MK-4 to MK-13), as well as a synthetic form, vitamin K3 menadione. The key function of vitamin K is to act as a cofactor for gamma-glutamyl carboxylase (GGCX), which is responsible for the biosynthesis of the γ-carboxyglutamate moiety of proteins using the redox cycle of vitamin K (Furie et al. 1999; Rishavy and Berkner 2012; Schurgers et al. 2008; Stafford 2005). GGCX is the only known carboxylase in humans and 14 substrate proteins have been identifeid. Carboxylation is the process of biosynthesis of gamma-carboxyglutamic acid moiety in vitamin-K-dependent (VKD) proteins, which confers the ability to chelate calcium ions, enabling these proteins to function as matrix coagulators and calcification preventors (Schurgers et al. 2013). As our previous studies have shown both proton and Ca^{2+} ions are involved in the vitamin K2 redox-cycle dependent carboxylation function of GGCX (Ma et al. 2019). This is consistent with the notion of interplay of acidity and Ca^{2+} in the epididymal luminal microenvironment, which is a tightly regulated physiological condition for sperm maturation (Shum et al. 2009, 2022). In the epididymis, our studies have also demonstrated that the VKD proteins, MGP and GGCX, are also localized to the EVs-like structures in the epididymal lumen and spermatozoa (Ma et al. 2019), consistent with the finding of others that MGP is loaded in the EVs the vasculature (Reynolds et al. 2004). Recently, a novel, carboxylation-independent, and potent ferroptosis suppressor function has been reported for vitamin K cycle (Mishima et al. 2022). In this study, complying with its redox property, vitamin K acts as an electron carrier for ATP production and prevents ferroptosis, a type of cell death hallmarked by lipid peroxidation, which is regulated by ferroptosis suppressor prtein-1 in a vitamin K-dependent manner. This is indeed in line with the fact that the crosstalk between EVs and ferroptosis has been an interesting biological question with promising perspects for further understanding the biology and potential development of therapeutics and diagnostics (Mishima et al. 2023; Wu et al. 2024).

4.4 *Water-Soluble Vitamins*

(i) *Vitamin B*

B vitamins, also known as B complex vitamins, are water-soluble vitamins that consist of eight members: thiamin (B1), riboflavin (B2), niacin (B3), pantothenic acid (B5), pyridoxine (B6), biotin (B7), folic acid (B9), and cobalamin (B12). Except vitamin B3, which can also be synthetized from amino acid tryptophan, all other B vitamins must be consumed regularly and obtained from diet. Since B vitamins are water soluble, they are not stored in excess amounts in the body and are excreted, so they are generally considered safe even when taken in high doses in supplements or through dietary sources (Hrubsa et al. 2022; Lyon et al. 2020). All B vitamins play essential roles as coenzymes or precursors to coenzymes, and thus they assist enzymes in facilitating metabolic reactions in the body (Franco et al. 2022; Hrubsa et al. 2022; Lyon et al. 2020). For example, the well-known function of B vitamins, including B1, B2, and B3, is its involvement in energy metabolism through redox reactions in the Krebs cycle. In addition, many B vitamins work synergistically, so deficiencies in one can affect others (Lyon et al. 2020). For example, several B vitamins—including B1, B2, B3, B6, B9, and B12—play essential roles, either directly or indirectly, in one-carbon metabolism, a critical biochemical network responsible for transferring methyl group (-CH_3) in DNA synthesis (e.g., purine and thymidylate synthesis), amino acid metabolism (e.g., homocysteine remethylation), and epigenetic regulation (Lyon et al. 2020). As the functions and biochemical properties of B vitamins have been thoroughly reviewed by others in non-reproductive contexts, we briefly review here the literature on their relevance to epididymal sperm maturation and the potential involvement of EVs in this regard.

(a) *Vitamin B1 (Thiamine).* In the epididymis, vitamin B1—primarily in its active form thiamin pyrophosphate—appears to play a particular important role by enhancing the capacity of rat cauda sperm to initiate motility following storage in native cauda lumen fluid (Turner 2002). This is partly due to the fact that vitamin B1 plays a crucial role in the conversion of pyruvate to acetyl-CoA, a key step in the energy production in mitochondria (Lyon et al. 2020), consistent with the fact that sperm use lactate and pyruvate as energy sources (Darr et al. 2016). The concentrations of thiamin are low in both blood plasma and seminal plasma of humans, which are only in the tens of nanomolar range (Gangolf et al. 2010). This may be due to the high instability of carbene, the functional molecule of thiamine, in aqueous conditions (Raviprolu et al. 2025). However, male mice lacking thiamine transporter *Slci9a2* are infertile (Fleming et al. 2003). Inclusion of vitamin B1 and B2 to the cryoprotectants has also showed protective effect in the qualities of sperm and seminal plasma (Tabatabaei Vakili et al. 2025). The protective function of these vitamins in sperm may involve their antioxidant effect and regulation of cellular calcium homeostasis (Huang et al. 2010). Although no thiamine and its derivatives are found in non-reproductive EVs or epididymosomes (Keerthikumar et al. 2016; Thimon et al. 2008), it has been reported that thiamine can be transported in vesicles in the rat renal brush border membrane (Gastaldi et al. 2000).

(b) *Vitamin B2 (Riboflavin).* Riboflavin is the precursor of the active forms of flavin adenine dinucleotide (FAD) and flavin mononucleotide (FMN) coenzymes, which act as electron carriers and participate in a number of redox reactions involved in energy production, cellular antioxidant function, and numerous metabolic pathways (Hrubsa et al. 2022; Lyon et al. 2020). In the male reproductive tract, a riboflavin transporter, *Slc52a3*, is detected in the mouse spermatozoa but not in the epididymal EVs (Skerget et al. 2015). In the non-reproductive EVs and bovine epididymosomes, the FAD to cyclic-FMN-forming lyase enzyme, *Tkfc*, has been detected (Girouard et al. 2009; Keerthikumar et al. 2016). Its optimal enzymatic activity occur at pH 6.6, a feature that is preserved within the epididymal lumen. This implies that vitamin B2 has an unexplored role in epididymal function. To support this speculation, our previous study also implies that vitamin B may be involved in the metabolism of intracellular MGP and may participate in the interplay of the homeodynamics in epididymis (Shum et al. 2022), although the underlying functional mechanism remains to be explored.

(c) *Vitamin B3 (Niacin).* The active forms of vitamin B3 are its derivative coenzymes, nicotinamide adenine dinucleotide (NAD), and nicotinamide adenine dinucleotide phosphate (NADP) (Lyon et al. 2020). They function as the electron carriers and play a critical role in redox reactions and energy metabolism, such as the catabolic and anabolic reactions occurring within the Krebs cycle. Deficiency of vitamin B3 leads to low levels of NAD/NADP, also commonly known as pellagra, which has implications for aging, metabolic dysfunction, and neurodegenerative diseases (Camacho-Pereira et al. 2016; Hrubsa et al. 2022; Tarrago et al. 2018;

Wu and Zhang 2020; Xie et al. 2020). The NAD precursors such as nicotinamide riboside and nicotinamide mononucleotide (NMN) have been used as supplements to support NAD levels and promote health (Covarrubias et al. 2021). NAD is thus considered a key factor in maintaining youthfulness. It is important to note that NAD is dynamically metabolized in cells and that a balance level of NAD through synthesis and degradation is crucial for maintaining cellular function and overall health, including longevity and age-related health benefits. And it seems that NAD level is not positively correlated with sperm quality. Administration of NAD-booster NMN can ameliorate the negative effects of obesity-induced poor sperm quality at appropriate dosage and administration approach (Youngson et al. 2019); however, unhealthy excessive levels of NAD and/or its metabolites is known to be harmful (Ferrell et al. 2024); increased NAD levels after excessive supplementation of NMN, the precursor of NAD, in the testis of mice have been linked to reductions in sperm count and vitality and increased sperm oxidative DNA damage (Youngson et al. 2019). And recently, it has been reported that NAD depletion does not corelate with aging, as revealed by skeletal muscle function (Chubanava et al. 2025). Unlike the age-related decrease of NAD levels in somatic tissues, the sperm NAD concentration is not age dependent and indeed is negatively correlated with sperm quality in men with oligozoospermia (Bai and Wang 2022). This is consistent with the report that the redox state of NAD/NADH is associated with sperm quality in idiopathic infertility (Galimov et al. 2021). In the epididymosomes of bovine and ovine, NADH-dependent and NADPH-dependent cytochrome reductases and NADP-dependent isocitrate dehydrogenase have been detected (Girouard et al. 2011), although their functional roles in epididymis are not fully clear yet. Thus, the functional mechanistic role of vitamin B3 in epididymal function, either with or without EVs, remains to be explored.

(d) *Vitamin B5 (Pantothenic acid).* As an essential vitamin, vitamin B5 pantothenic acid does not participate in redox reactions directly, instead it acts as a precursor molecule in the synthesis of coenzyme A (CoA), a crucial coenzyme for many acyl-group dependent redox-sensitive metabolic pathways, including energy metabolism (mitochondrial metabolism), fatty acid synthesis, as well as sex- and stress-related hormones generation (Hrubsa et al. 2022; Lyon et al. 2020). The free acid form of vitamin B5 is stable at neutral pH, but becomes unstable on exposure to moisture, heat, acid, and alkali. It can be stabilized by forming complex with calcium ions in a deprotonated form (Corinti et al. 2021). It can also form double salt with isobutyrate or lactate molecules (Liang et al. 2021). These biochemical properties suggest that the functional role of vitamin B5 is sensitive to the change of acidity and calcium concentration. Interestingly, our previous studies also preliminarily revealed a potential involvement of B vitamins in vitamin K-dependent calcium homeostasis in the epididymis, an important process that is associated with asthenozoospermia risk. This may occur when the acid-promoted, redox-sensitive vitamin-K cycle is interupted by inhibition of the carboxylation enzyme GGCX (Ma et al. 2019; Shum et al. 2022). Vitamin B5 itself has not been directly detected in epididymal EVs. However, its derivative, pantethine, has been reported to inhibit the biogenesis of non-reproductive EVs (Kavian et al. 2015; Penet et al. 2008), presumably through its involvement in lipid metabolism and its capacity to reduce total cholesterol levels in cellular membranes (Ranganathan et al. 1982).

(e) *Vitamin B6 (Pyridoxine).* The B6 vitamin comprises a group of six chemically related compounds cored with a pyridine ring. Although B6 vitamins are not directly involved in redox reactions, their Schiff base chemistry and the electron sink properties make them highly versatile coenzymes. They serve as cofactors for over 150 enzymes, primarily involved in amino acid synthesis and degradation (Lyon et al. 2020; Parra et al. 2018). Vitamin B6 is particularly essential for tryptophan metabolism—a precursor pathway for NAD production—as well as for lipid metabolism. Vitamin B6 deficiency affects a range of physiological processes and may lead to serious complication such as epilepsy (Parra et al. 2018). Vitamin B6 toxicity by overdose from diet is unlikely but it may happen by over-supplementation. High doses of vitamin B6 have been shown to impair testicular spermatogenesis (Kaido et al. 1991). The physiological ranges of vitamin B6 showed no correlation with sperm motility in normal men compared to men with asthenozoospermia (Banihani and Aljabali 2020). Our previous bioinformatic analysis reveals potential involvement of vitamin B6 pathways in vitamin K-dependent GGCX in regulating the extracellular chelating clearance function of MGP, which is under Ca^{2+} modulation, thereby modulating Ca^{2+} homeostasis in the epididymis (Shum et al. 2022). Since B vitamins operate synergistically with redox-dependent energy metabolism, it is speculated that the metabolism of carboxylation-decarboxylation cycle of MGP may be involved in this process. Given that MGP is loaded in vascular EVs and plays an important role in regulation of vascular calcification process (Kapustin et al. 2011, 2015; Reynolds et al. 2004), the involvement of epididymal EVs in a similar context remains to be investigated.

(f) *Vitamin B7 (Biotin).* Biotin is a small molecule with a ring containing nitrogen and carbonyl groups and a sulfur-containing heterocyclic ring, as well as a carboxylic acid-terminated hydrocarbon chain. Biotin plays an essential role in carboxylation reactions, and because of this, it also has many applications in laboratory research (Lanska 2012; Waldrop et al. 2012). The carboxylase domains of biotin-dependent proteins are often referred to as "two metal enzymes" because they require two metal ions for their catalytic activities (Waldrop et al. 2012). In these two-metal biotin-dependent carboxylase proteins, magnesium (Mg^{2+}) functions as primary catalytic metal and potassium (K^{+}) or manganese (Mn^{2+}) as secondary metal that stabilize the transition state. Consistent with its coenzyme role as the metal enzymes of carboxylation, biotin B7 can prevent magnesium-induced toxicity (Lai et al. 2023). It has been reported that biotin sperm cryopreservation medium can improve sperm survival without affecting the fertilizing ability, involving protection of DNA and reduction in protein carbonyl content, and preservation of sperm plasma membrane and acrosome integrity (Ranjan et al. 2022). In parallel, biotination techniques have been widely applied in EV research to enable more detailed profiling of surface proteins. Using these methods, studies on prostate-derived EVs and seminal fluid EVs have identified previously unreported proteins, such as GAPDS and PTGDS (Ranjan et al. 2022). It is expected that similar approaches can be employed in the future for more in-depth investigations of epididymal EVs and epididymosomes to reveal under-recognized surface proteins.

(g) *Vitamin B9 (Folic acid).* Vitamin B9 is a generic term that refers to a group of several molecules with a folic acid structure, including mainly folate that occurs naturally and folic acid that is commonly used in supplements in synthetic form. The molecule of B9 vitamin contains three core structures: a nitrogen-rich aromatic pteridine ring, a para-aminobenzoic acid, and polyglutamic acid, which can be liked in ϒ-bond by 2–7 residues (Siatka et al. 2025). Studies have shown that the polyglutamate tail increases the cellular retention of vitamin B9 and significantly increases the affinity of folate acid for folate-dependent enzymes (Siatka et al. 2025). B9 mediates one-carbon (1C) metabolism and involves in many metabolic processes, such as catabolizing of serine 3-carbon to 1C metabolism in cellular cytosols and glycine degradation in mitochondria (Lyon et al. 2020; Tan et al. 2020). It is well known that vitamin B9 deficiency can lead to adverse pregnancy outcomes, but this is not only due to malnutrition in women but also due to poor nutrition that impacts reproductive health in men (Wallock et al. 2001). Several forms of folate, including 5-methyltetrahydrofolate and non-methyltetrahydrofolates, are present in seminal plasma, with total concentrations about 1.5-fold higher than in blood plasma. Folate status appears to be associated with sperm count and density, and a total and 5-methyltetrahydrofolate concentrates correlate with folate status with sperm count and density of non-smokers men, and non-methyltetrahydrofolate concentration is associated inversely in smokers (Wallock et al. 2001). A meta-analysis of studies shows that paternal folate concentrations and their status in humans and animals have positive impact on sperm quality and pregnancy outcomes, including sperm DNA integrity, DNA methylation, placentation, embryonic growth, and development (Hoek et al. 2020). Folate seems to function as free molecule in male reproductive fluids, although the investigation of folate status and concentration in epididymal fluid is lacking. A folate-related protein, cytosolic C-1-tetrahydrofolate synthase (*Mthfd1*), also known as epididymis secretory sperm binding protein, was detected in non-reproductive EVs (Keerthikumar et al. 2016) and mouse spermatozoa

(Skerget et al. 2015), but not in epididymosomes. In plant EVs derived from ginger, folic acid can be conjugated to the EVs to enrich their folate content in supplements. These engineered EVs have been used for targeted treatment of rheumatoid arthritis by modeling microenvironment, potentially through involvement of the PI3K-AKT pathway (Han et al. 2025). This implies that this approach of supplementing optimal dosage of vitamin B6 by conjugating to epididymal EVs could be an alternative targeted intervention to improving spermatozoa quality and male reproductive outcomes.

(h) *Vitamin B12 (Cobalamin).* Vitamin B12 is a group of essential biomolecules that are uniquely based on their core structure with stable metal-carbon bond, namely the complex organometallic compounds, or cobalamins. These compounds all have a corrin ring as its core structure, with the essential element cobalt ion (Co) located at the center of the corrin ring, sandwiched between the upper ligand at β-position and the lower ligand at α-position (Dereven'kov et al. 2013; Randaccio et al. 2010). Vitamin B12 functions as a redox-active cofactor, involving cycling redox between cobalt metal oxidation states, existing in +3, +2, and + 1 forms, and participates in a variety of reactions as antioxidant and electron donor. The ligand of Co ion determines the form and function of B12 in the body. The oxidation state of the cobalt ion in cobalamins directly influences the compound's coordination and redox behavior, and the ligand binding is subjected to protonation of cobalamins and thus under the influence of acid-base balance (Dereven'kov et al. 2013; Randaccio et al. 2010). Vitamin B12 is essential for DNA synthesis, taurine-synthesis, methylation reactions, energy production, and redox metabolism (Offringa et al. 2021; Randaccio et al. 2010; Roman-Garcia et al. 2014; Schumacher et al. 1997; Suarez-Moreira et al. 2009; van de Lagemaat et al. 2019). All these functional roles of vitamin B12 are essential for both male and female reproductive health, but the specific functional mechanisms of B12 are still largely unclear (Pront et al. 2009). Deficiency of vitamin B12 can be associated with chronic genital pain in men (Cui and Terlecki 2018) and semen quality (Banihani 2017), and its supplementation showed beneficial effects on post-thaw human sperm (Hosseinabadi et al. 2020). B12 has also been shown to mitigate the adverse consequences on sperm quality in patients treated with cancer therapeutic cimetidine, which is associated with androgen disruption (Beltrame et al. 2019). The concentration of vitamin B12 transport protein, transcobalamin II (TCN2), also known as beta-globulin, is the highest in human seminal plasma, about ten times that of plasma, and the seminal vesicle is believed to be the site of its production (Carmel and Bernstein 1984). Lower concentration of cobalamin was found in azoospermic seminal plasma than normozoospermic, whereas cobalamin (and folate) was higher in seminal plasma from obstructive azoospermia than in non-obstructive azoospermia (Crha et al. 2010). The presence of cobalamin-binding proteins including TCN2 in the secretion of kidney epithelial cells has been reported (Brada et al. 2000), an organ that shares many physiological aspects similar to epididymis; thus, the presence of TCN2 in epididymal epithelial cells is awaited to be validated. A cobalamin-related protein, lysosomal cobalamin transport escort protein (LMBD1), was detected in non-reproductive EVs (Keerthikumar et al. 2016), but not in spermatozoa or epididymosomes. It seems vitamin B12 nutrients can be smuggled through gut bacteria-derived EVs as they have been shown to scavenge vitamin B12 and related derivatives for the commensal microbe in the gut and host acquisition through intestinal epithelial absorption (Juodeikis et al. 2022). This is consistent with the notion that intestinal microbe serves as a source of essential nutrients to support male reproduction (Cai et al. 2022). The role of EVs in the circulation and epididymal luminal microenvironment remains to be explored.

(ii) *Vitamin C.*

Vitamin C is water-soluble ROS scavenger with high potency. It exists in two primary active forms in humans: the reduced form (ascorbic acid) and the oxidized form (dehydroascorbic acid, DHA) (Shen et al. 2021). Ascorbic acid occurs naturally as a potent antioxidant and acts directly as electron donor. DHA is now known as indirect antioxidant as it does not donate electron directly under physiological conditions (Diaz et al. 2021; Shen et al. 2021). Vitamin C does not appear to have pro-oxidant properties in vivo (Carr and Frei 1999), because the redox-active metals, copper (Cu) and iron (Fe), act cooperatively with it as redox catalysts with preference even over free radicals (Shen et al. 2021). The oxidized form of vitamin C, DHA, can be converted back to ascorbic acid readily in the body. It can also cross some cell membranes and blood-tissue barrier more easily than ascorbic acid. The benefit of antioxidant effect of vitamin C is well documented (Colagar and Marzony 2009; Das et al. 2009; Menditto et al. 1997; Rolf et al. 1999; Sonmez et al. 2005). In the male reproductive tract, the concentrations of ascorbic acid in the epididymal fluid and seminal fluid are in millimolar range, which are almost tenfold higher than those in blood plasma (Jacob et al. 1992). In the semen of men with idiopathic infertility, the ascorbic acid content is significantly lower than that in men with normal fertility, and the concentration in the latter is lower than that in the epididymal fluid, but still much higher than that in blood plasma (Colagar and Marzony 2009). A daily supplementation of vitamin C in the range of 1,000 mg per day is therefore recommended (Rolf et al. 1999). However, the adverse effect of reductive stress is paradoxically induced by overnutrition with multi-vitamin supplementation (Zhang et al. 2025b). It is speculated that high doses of antioxidants such as vitamin C might accelerate metabolic disorders by preventing the alleviation of insulin resistance (Zhang et al. 2025b). Thus, while the dosage in situations of overnutrition and when combined with other micronutrients is unclear (Hondal 2023), the use of high-dose antioxidants as supplements in men seeking for reproductive assistance should be approached with caution, especially the physiological impacts of reductive stress on epididymal sperm maturation remain unclear. It has been reported that vitamin C can exert its redox ability to break the disulfide bonds in sperm protamines, leading to asynchronous chromosomal decondensation and reduced fertility (Henkel et al. 2019). Given that the formation of disulfide bonds in protamines requires a moderate oxidation energy (Oliva 2006), one may confer that it is the reductive stress of vitamin C that interrupts the energy and leads to the disassociation of disulfide bonds. The dietary source of ascorbic acid from EVs has been documented to be of plant origin (Jiang et al. 2024a; Mu et al. 2023; Perut et al. 2021). In mammalian system, ascorbic acid can be transferred from astrocytes to neurons as a free molecule or via EVs; however, these different routes of ascorbic acid display distinct functional roles (Beltran et al. 2025). This study reports that vitamin C-containing EVs effectively reduce the ROS, whereas free vitamin C is involved in regulating neuronal metabolic substrate preference; moreover, the release of vitamin C-containing EVs is interrupted in the neurodegenerative order of Huntington's disease. In the epididymis, whether ascorbic acid similarly functions via EV-mediated pathways to support sperm maturation—and the underlying mechanism—remains to be elucidated.

(iii) *Ergothioneine*

Ergothioneine is a thiol-histidine derivative amino acid with potent antioxidant activity. It is known as a stress vitamin and is shown to be beneficial for anti-aging and longevity and may have neuroprotective functions (Ames 2018; Paul 2022). Ergothioneine is synthesized by microorganisms and only absorbed by human tissues through diet via specific absorption process. Ergothioneine was discovered in 1909, and as early as in 1950s, it was found in seminal vesicular secretion (Leone and Mann 1951) and then confirmed to have a counteracting effect on sperm inhibition induced by peroxidase content in uterine fluid (Smith and Klebanoff 1970). It has then been used as an antioxidant in semen and sperm cryopreservation (Kaewma et al. 2024; Najafi et al. 2014; Younus et al. 2024). Recently, ergothioneine has been reported to control mitochondrial function by directly activating 3-mercaptopyruvate sulfur transferase, thereby promoting cellular respiration (Sprenger et al. 2025). Regarding its role in EVs, although no ergothioneine is directly detected in EVs, in rats with varicocele-induced testicular damage (Chen et al. 2023), ergothioneine can upregulate the levels of HSP90AA1, a known marker of EVs including epididymosomes (Keerthikumar et al. 2016; Thimon et al. 2008), suggesting that it plays a functional role via epididymal EVs in promoting sperm maturation.

4.5 Mineral Nutrients

Mineral nutrients are essential inorganic elements for human health. These essential minerals, both metals and nonmetals, cannot be synthesized by the body and must be obtained through diet or supplementation in ionic forms, or in organic form, such as selenomethionine. Two primary categories are classified for mineral nutrients based on their daily requirement in humans: macrominerals that are needed in relatively larger amounts and microminerals that are also known as trace elements and needed in smaller amounts (Mirnamniha et al. 2019; Tvrda et al. 2015; Zoroddu et al. 2019). Macrominerals include metallic calcium (Ca), magnesium (Mg), potassium (K), sodium (Na), and non-metallic chlorine (Cl), fluoride (F), hydrogen (H), nitrogen (N), oxygen (O), sulfur (S). Their roles in epididymis and sperm maturation involve various physiological processes, such as structural and regulatory functions, which have also been relatively well studied and reviewed and are therefore not the focus of this discussion (Frizzell and Hanrahan 2012; Gervasi and Visconti 2017; Robaire and Hinton 2015; Shum et al. 2009, 2022; Wong et al. 2002; Zhou et al. 2018). Microminerals are required in trace amount but critical for enzyme systems, hormones, and metabolic pathways and include metals such as cobalt (Co), copper (Cu), iron (Fe), manganese (Mn), and zinc (Zn) and nonmetals such as selenium (Se) and iodine (I). Nearly all the microminerals, except for Zn, directly mediate redox reactions within their elemental cores, i.e., cycles of electron donation and electron acceptance. It can be conferred that the redox metabolism of these trance minerals is intrinsically linked to the redox metabolism of epididymal epithelial cells and spermatozoa and thus the functional metabolism of the entire organ. However, their roles in epididymis are still largely unexplored. This section will discuss the literature on the association of microminerals with epididymal sperm maturation based on their redox sensitivity, although most of their functional roles remain largely unclear.

4.6 Redox-Sensitive Metal Microminerals

(i) *Cobalt (Co)*

Cobalt is an essential trace mineral because it serves as the central component of vitamin B12 (commonly known as cobalamins); however, the human body does not utilize free cobalt independently. The function of cobalt in vitamin B12 involves redox cycling between its +3, +2 and +1 oxidation states. This redox flexibility directly influences the structural form, redox behavior, and ultimately the biological function of vitamin B12 (Dereven'kov et al. 2013). Cobalt in vitamin B12 with oxidation state +1 is in its reduced active form (Cbl(I)), which participates in methyltransferase reactions (e.g., methionine synthase). Cobalt in +2-oxidation state (Cbl(II)) is a semi-stable oxidized intermediate, while cobalt in +3-oxidation state (Cbl(III)) represents the oxidized from. All these forms are commonly found in supplements. Although the oxidized forms (Cbl(II)) and (Cbl(III)) mat transiently exist in the body, they are reduced back to the active (Cbl(I)) in order to carry out their biological functions. The redox metabolism of oxidized cobalamins to their reduced form involves reducing enzymes, including methionine synthase reductase (MTRR) that uses S-adnosylmethionine (SAM) and NADPH as cofactors, GSH and TRX system, and FMN/FAD-dependent enzymes that besides uses the NADH-dependent reductases but also indirectly involves cytochrome P450 reductase (Offringa et al. 2021). Some of these B12 reducing enzymes are known to be expressed in the epididymis and play important roles in sperm maturation and male reproduction, including epididymosomes (Girouard et al. 2011; Rinaldi et al. 2020; Shi et al. 2021). The involvement of epididymal EVs may depend on the same or other enzymes as discussed above. Further research is required to fully understand the redox metabolism of cobalamins in epididymal sperm maturation and male reproduction.

(ii) *Copper (Cu)*

Copper (Cu) is a transition metal and one of the redox active essential trace minerals that plays a critical role in various physiological processes in the human body. Key functions of copper, just to name a few, include energy production, iron metabolism, antioxidant defense, neurotransmitter synthesis enhancement, connective tissue health, gene regulation and epigenetics, and enzyme cofactor (Michniewicz et al. 2021; Tsang et al. 2021). The function of copper depends on its unique biochemical property, particularly its ability to undergo single-electron transferring redox activity from its reduced Cu^{+1} state to oxidized Cu^{+2} state, a process readily occurs in the body. In this redox cycle, the oxidized divalent copper can oxidize superoxide anions in cells into oxygen through the Haber-Weiss reaction, while itself being reduced to monovalent copper. The monovalent copper can then reduce the hydrogen peroxide in the cell into more cytotoxic hydroxyl free radicals by participating in the Fenton reaction, which can then be oxidized back into divalent copper. In addition to redox activity, copper also has high binding affinity for sulfur- and nitrogen-containing moieties in organic matters, such as thiol groups in cysteine and glutathione, and imidazole nitrogen in histidine. This binding property can influence the other functional properties of copper, including chemical coordination in proteins and enzymatic catalytic mechanisms, as well as the bioavailability of copper in specific biological compartments (Chen et al. 2022; Culotta 2010; Ogorek et al. 2019; Tsang et al. 2021). One of the well-known mechanistic functions of copper is its association with cell death process, which is now called coprotopsis (Tsvetkov et al. 2022), a copper-dependent form of apoptosis, involving binding of copper to lipoylated enzymes of the TCA cycle, leading to protein aggregation, proteotoxic stress, and then cell death (Chen et al. 2022). In this process, the copper binding proteins, reductases (e.g., STEAP) and plasma membrane transporters (SLC31A1, ATP7A/B) are involved. Extracellular copper ionophore such as elesclomol is involved in trafficking of copper (Chen et al. 2022; Yang et al. 2024). Maintaining copper homeostasis is also important to maintain iron metabolism, particularly by involving LOX and ceruloplasmin function (Yang et al.

2024). In the epididymal fluid, abundant ceruloplasmin was present in the epididymal fluid and believed to participate in copper metabolism by transporting copper in the extracellular lumen (Ogorek et al. 2019). The copper ion transporter CTR1 is expressed on the apical and basal surfaces of the principal cells of epididymal epithelium (Ogorek et al. 2019). The transmembrane copper transporter ATP7B is mainly located at the apical surface of epididymal principal cells, whereas ATP7A is on the basal surface of principal cells as well as basal cells in the caput region (Ogorek et al. 2019). It has been known that copper makes it a double-edged sword in maintaining redox homeostasis (Tvrda et al. 2015). Copper deficiency can lead to higher levels of ROS, iron toxicity, and downregulation of enzymatic activity, and eventually can result in impaired sperm parameter and male fertility. When copper is in excess in the body, which can be caused by a rare inherited condition of Wilson's disease, copper toxicity occurs due to abnormal accumulation of copper in serveral organs (Chen et al. 2022; Yang et al. 2024), including male reproductive organ (Mirnamniha et al. 2019). Chelatin therapy is used for the treatment of copper toxicity. Targeting the copper uptake process involving the CD44 surface glycoprotein with the novel copper binding compound LCC-12 and regulating intracellular copper homeostasis have emerged as a promising target for the treatment of diseases associated with copper overload (Solier et al. 2023). As for the potential role of epididymal EVs in copper metabolism, although ATP7A and ceruloplasmin have been detected in non-reproductive EVs, they are only found in epididymal spermatozoa but not in epididymal EVs (Girouard et al. 2011; Keerthikumar et al. 2016). More studies are required to further dissect the biology of copper metabolism involving other proteins in the epididymal luminal microenvironment and sperm maturation.

(iii) *Iron (Fe)*

Iron (Fe) is also an essential mineral and plays several critical roles in the human body. Besides its role in oxygen transport as a carrier and energy production as a cofactor, it also participates in redox biology, immunity, cognition function, DNA synthesis, detoxification, as well as thyroid metabolism (Abbaspour et al. 2014; Arredondo and Nunez 2005). In order to absorb iron from the nature, organisms including humans develop iron-binding complex forms, such as hemoglobin, heme enzymes, or nonheme compounds (transferrin and ferritin). Deficiency of iron intake and/or low bioavailability lead to anemia. In the body, iron is bound and transproted via transferrin and stored in ferritin molecules. Once absorbed, there is no physiological mechanism for excretion of excess iron from the body other than blood loss (i.e., pregnancy, menstruation, or other bleeding). Iron overload can lead to hemochromatosis, which is more susceptible in men than women, and is a result of insufficient hepcidin levels (Ganz and Nemeth 2012). Interestingly, the expression of hepcidin, a liver hormone controlling iron absorption and hemeostasis, is regulated by acidity (Mizumoto et al. 2012) and involves endogenous BMP6, which acts as a ligand of soluble hemojuvelin (Andriopoulos Jr. et al. 2009). The activity of iron depends on its ability to switch between ferrous (Fe^{2+}) and ferrric (Fe^{3+}) states in the body. For example, iron bound in heme is in the Fe^{2+} state, and it exists in the Fe^{3+}

state when bound to transferrin or functioning in dotoxificying enzymes such as catalase and peroxidase; in the cytochorme enzymes, it can alternate between Fe^{2+} and Fe^{3+} states as an electron carrier (Abbaspour et al. 2014; Arredondo and Nunez 2005). The redox metabolism of iron is closely related to copper metabolism in the human body, and their uptake process in cells is competitive in nature and is influenced by the presence of vitamins such as vitamin C (Arredondo and Nunez 2005). One core iron-dependent physiological process is ferroptosis, a form of programmed cell death characterized by the accumulation of lipid hydroperoxides. Due to the critical role of iron in redox reactions in the body, ferroptosis is also considered a center process of redox mebolism and is closely associated with various diseases (Stockwell et al. 2017). In addition to vitamin C, a new non-canonical vitamin-K-dependent redox cycle regulator, ferroptosis suppressor protein 1 (FSP1) is found to be a potent ferroptosis suppressor (Mishima et al. 2022). In a recent study, FSP1 is discovered to be vitamin K reductase that overcomes warfarin poinsoing and protects against ferroptosis (Li et al. 2023), and involves VKORC1L1 (Wang 2023). In the epididymal EVs and on the spermatozoa of corpus and cauda regions, an iron binding and transporting protein, lactotransferrin (LTF), was detected (Gatti et al. 2005; Girouard et al. 2011; Keerthikumar et al. 2016; Skerget et al. 2015; Thimon et al. 2008), which is belived to participate in innate immune defence and antimicrobial functions in the extracellular fluids (González-Chávez et al. 2009). Currently, there is little research on the biological and functional mechanisms of iorn in sperm maturation in the epididymis, and further exploration is needed.

(iv) *Manganese (Mn)*

Manganese (Mn) is one of the essential trace minerals in the body and is required in small amount. It acts as as redox mediator in antioxidant defense, and as a cofactor for critical enzymes, including glutamine synthetase (GS), pyruvate carboxylase, and mitochondrial enzyme Mn-SOD (or SOD2). The latter suggests that magnanese is required for enzymes of almost all cells in the human body that carry out aerobic respiration. Through these metaloproteins and its redox and binding capacity, managanese plays an indispensable roles in processes including antioxidant defense, development, digestion, energy production, immune response, metabolism, reproduction, regulation of neuronal activities, wound healing (Avila et al. 2013; Chen et al. 2018). Manganese difficiency rarely occurs, but it is possible and may lead to weak bones, poor glucose tolerance, and altered lipid metabolism. On the contrary, magnanese toxicity may occur upon overexposure to the mineral, which leads to oxidative stress, mitochondrial dysfunction and other cellular stress, and disruptive imbalances in other micronutrients such as vitamin K and other redox-active minerals (Aschner and Aschner 2005; Baj et al. 2023; Chen et al. 2018). It is known thant manganese uptake is controlled by Cu-dependent ferroxidase ceruloplasmin (Jursa and Smith 2009), a protein that is abundantly present in the epididymal luminal fluid (Ogorek et al. 2019). It has been reported that ceruloplasmin plays a significant role in the transport of maganese to tissues and contributes to the complex interactions between metal oxidases and toxicity of redox labile metals such as Cu and Fe. Mn-SOD is a key enzyme that used manganese-containing enzyme as a cofator in the mitochondrial matrix to convert harmful ROS into H_2O_2 (Zelko et al. 2002). Mn-SOD is a member of superoxide dismutase family that is vital for life, as reflected by the fact that genetic disruption of Mn-SOD is lethal in mice (Lebovitz et al. 1996). Another member of this superoxide dismutase family is CuZn-SOD, which is found in the cytoplasm of cells and works as a primary antioxidative enzyme scavenging ROS (Lebovitz et al. 1996). Both Mn-SOD and CuZn-SOD play important roles in sperm and epididymis functions (Garratt et al. 2013; Nonogaki et al. 1992; Studer et al. 2022; Vernet et al. 2004). In addition to the mitochondrial Mn-SOD and cytosolic CuZn-SOD, secreted form of CuZn-SOD is found in exosomes (Keerthikumar et al. 2016; Sentman et al. 2006). In the epididymis, CuZn-SOD is consistently found in the EVs of bovine and ovine but not human epididymosomes, as well as in mouse spermatozoa (Gatti et al. 2005; Girouard et al. 2009; Thimon et al. 2008). The secreted form of dismutase is believed to catalyze the dismutation of superoxide to H_2O_2 in the luminal microenvironment. Although manganese-bound ceruloplasmin has been detected in non-reproductive EVs, it has only been found in epididymal spermatozoa and not in the epididymal EVs (Girouard et al. 2011; Keerthikumar et al. 2016). This suggests a potential intriguiling intercellular interplay involving redox-labile minnerals in the epididymal lumen, which warrants further investigation.

(v) *Molybdenum (Mo)*

Molybdenum (Mo) is an essential trace element for the human body and plays a critical role in redox reactions in all organisms (Hille et al. 2014; Mendel and Kruse 2012; Wang 2012). Molybdenum primarily acts as a cofactor for key enzymes involved in sulfur metabolism, detoxification, and antioxidant defense. Molybdoenzymes require molybdenum as a cofactor to activate and are important for some crucial redox reactions, which involve xanthine oxidase (XO), aldehyde oxidase (AO), sulfite oxidase (SO), and/or mitochondrial amidoxime reducing component (mARC) (Wahl et al. 2010). The cofactor function of molybdenum relies on its binding ability to another cofactor, pterin, that is located in the active center of all molybdenum-dependent enzymes (Hille et al. 2014). Deficiency in molybdenum cofactor can lead to severe metabolic defect. This is particularly related to a condition known as molybdenum cofctor deficiency (MoCD), a rare autosomal recessive inborn error of metabolism due to mutations of genes involved in MoCD biosynthesis (Misko et al. 1993; Reiss and Hahnewald 2011). The common symptoms of MoCD include neonatal seizures, facial dysmorphism, and microcephaly, and low serum uric acid and elevated urinary sulfite levels are found to be associated with MoCD (Reiss and Hahnewald 2011). Excessive molybdenum may lead to guotlike symptoms due to elevated uric acid, copper deficiency due to competetion absorption, and testicular atrophy and sperm quality (Briffa et al. 2020; Zhai et al. 2013). The excretion mechanism of molybdenum has not been characterized yet (Briffa et al. 2020; Mendel and Kruse 2012; Schwarz 2016). Toxicity of molybdenum may also be synergized when other toxic elements (e.g., cadmium) are present, and in such scenario, they coinduce pyroptosis and autophagy via ROS/PTEN/PI3K/

AKT axis in kidney epithelial cells, a cellular system that shares many similar aspects with epididymis (Zhang et al. 2021a). The effect of molybdenum on male reproduction appears to depend on the dose and the individual's overall health and has a dual effect. At low levels, it enhances sperm quality by regulatng oxidative stress in the testis and in the epididymis; thus, it can improve male fertility by repairing testicular damage and normalizing hormone levels. High levels of molybdenum exposure can negatively impact sperm quality, motility, and testicular function (Briffa et al. 2020; Zhai et al. 2013). The toxicology of co-exposure of molybdenum and cadmium may involve diruption of the blood-testis barrier dysfunction through ROS-mediated NLRP3 inflammazome activation (Jiang et al. 2024b). In the macrophages, the participation of EVs derived from ginger rhizomes in inhibition of NLRP3 pyrotopsis activation has been reported (Chen et al. 2019). The specific functional roles of molybdenum in the epididymal function is thought to be associatd with molybdenum-dependent oxidases and proteins (XO, AO, SO and/or mARC), but their mechanism of actions and regulations, through epididymal EVs or not, are not fully understood and require further research.

(vi) *Zinc (Zn)*

Zinc (Zn) is an essential trace mineral that plays a critical role in numerous biological processes, despite not participating directly in redox reactions like some other transition metals (e.g., Cu, Fe, Mn). Instead, zinc indirectly regulates redox reactions and other cellular functions through its unique chemical properties and binding interactions. Zinc is the most abundant essential d-block metal ion in the body, second only to iron. However, in terms of enzyme usage as a cofactor, zinc has even higher proportion of interaction proteins (9%) than iorn (8%), following macromineral magnesium (16%) (Waldron et al. 2009). Zinc exists in the human body, and in all known biological systems, exclusively with oxidation state +2, or Zn^{2+}. It is redox stable and this property allows it to play sctructural and catalytic roles without contributing to oxidative stress, unlike the redox labile metals. Thus, zinc can be considered a major regulatory ion among the redox inert metal ions Na^+, K^+, Mg^{2+}, Ca^{2+}. Zinc acts as a structural or catalytic component in many proteins, including ZnCu-SOD, in which zinc stablilizes the enzyme' structure while copper handles the redox reaction. Zn^{2+} ions are able to behave as a second messenger, both intracellularly and extracellularly, a property it shares with calcium, and its regulatory network interacts with calcium signaling. The redox-inert nature of zinc makes it a better Lewis acid that stabilizes chromatin structure and DNA repair enzymes, indirectly protecting genomic integrity from oxidative damage (Zoroddu et al. 2019). In the epididymis, it is known that zinc is involved in the maturation of sperm, protecting them from oxidative damage during transit through the long journey down the epididymal tubule. Zinc deficiency can lead to poor sperm quality, reduced sperm count, and male infertility (Marin de Jesus et al. 2024). Zinc is transported by a network of proteins with the involvement of two main families of transporters, SLC30 (ZnT) for efflux and SLC39 (ZIP) for influx in cells, as well as the metallothioneins (MTs) for zinc storage and detoxification (Chen et al. 2024). In brain vesicles, in bones and muscles, high concentrations of zinc ions have been found (Zoroddu et al. 2019). In rat epididymis, zinc ions are also present in large quantities in vesicular deposits on the apical surface of principal cells and are more abundant in the cauda region than in the caput region (Stoltenberg et al. 1996). Consistent with the main localization on the cell membrane, the zinc transporter SLC30A7 has been detected in the bovine epididymosomes and non-reproductive EVs (Keerthikumar et al. 2016), as well as on the spermatozoa from mouse caput and corpus regions, but not on spermatozoa from cauda region (Gatti et al. 2005; Skerget et al. 2015). Although validation is requried, it has been suggested that epididymosomal zinc transporter SLC30A7 acts at the cell membrane and is invovled in the dynamic changes of sperm zinc content in different regions of epididymis. Targeting zinc transporters at the cell membrane appears to be a promising strategy for potential therapies and interventions to improve sperm quality and deserves further exploration.

4.7 Redox-Sensitive Nonmetal Microminerals

(i) *Selenium (Se)*

In humans, selenium is vital for antioxidant protection, thyroid function, immunity, and metabolism, as well as detoxification of heavy metals. The biological functions of selenium are related to its oxidation states, which can be at −2, 0, +4, and +6. Selenium in the oxidation state −2 is commonly found in the moiety of methylated selenium (CH_3Se^-) or in the selenoproteins, the latter containing the amino acids selenocysteine (SeCys) and/or selenomethionine (SeMet) (Kieliszek et al. 2021). In this reduced state, selenium serves as a key component of antioxidant enzymes, such as selenoprotein GPX, where it helps neutralize ROS and prevent ferroptosis (Ingold et al. 2018). Selenium also serves as a selenoenzyme in thyroid biosynthesis and in arsenic detoxication, where its methylated forms facilitate excretion via glutathione-dependent pathways (Kieliszek et al. 2021; Minich 2022; Sun et al. 2014a). Selenium in the oxidation state 0 is rare, but can be used as a detoxification process for excess selenium (Minich 2022). Selenium in the oxidation state +4 or + 6 is usually found in inorganic forms in nutritional sources and has pro-oxidant effects in high doses and can even causes toxic consequence when moderate doses are repeatedly ingested (Hadrup and Ravn-Haren 2023). Due to their effects on inducting oxidative stress, inorganic forms of selenium are used in cancer therapy (Kieliszek et al. 2021; Minich 2022). The various redox status of selenium that can occur in the body and the associated regulating networks suggest that both excessive and insufficient levels of this element have various complex impacts on human health (Hawkes and Turek 2001; Ingold et al. 2018; Maiorino et al. 2003). This is particular in the unique redox microenvironment for epididymal sperm maturation compared with circulating body fluids. It has been reported that both selenium deficiency and excess selenium may lead to abnormal sperm parameters and fertility disorders. Selenium improves sperm motility and vitality, which increases fertility outcomes and protection of sperm

DNA against oxidative stress (Hawkes and Turek 2001; Lakoma et al. 2023). However, in healthy men, high dosage of selenium derived from dietary supplementation resulted in decreased sperm motility and altered thyroid hormone metabolism; in addition, the change in selenium concentration was found only in seminal plasma but not in sperm (Hawkes and Turek 2001). Excessive selenium can be removed from the body on the expenses of methylation via one-carbon metabolism process by involving TRDs (Sun et al. 2014a), which plays an important role in the antioxidant defense system in the epididymis (O'Flaherty 2019). Selenoproteins such as GPX5 and TRX have been found in non-reproductive EVs and epididymosomes (Girouard et al. 2011; Keerthikumar et al. 2016), suggesting that selenium-mediated functions can be exerted through EVs. The functional applications and mechanisms of selenium in EVs on epididymal function and sperm maturation still need further investigations. Better intervention evaluation and strategies are needed to improve sperm quality and fertility outcomes using selenium supplementation, especially when combined with other vitamin supplements (Hondal 2023).

(ii) *Iodine (I)*

Iodine is an indispensable micronutrient for the biosynthesis of thyroid hormones, primarily thyroxin (T4) and triiodothyronine (T3), which are essential for regulating metabolism, growth, development, and reproductive functions. Iodine intake from the diet is uniquely concentrated in the thyroid gland where iodine serves as the key raw material in the production of thyroid hormones, by involving two key proteins: thyroid peroxidase (TPO) and thyroid hormone precursor thyroglobulin (Tg). Iodine found in the body and in the diet has an oxidation state of -1 in the most common active form (I^-). Whereas elemental iodine (I_2), in the oxidation state of 0, is generated during thyroid hormone synthesis through the oxidation of iodine I^- by TPO—a process that involves the reduction of H_2O_2, which is produced by NADPH oxidase at the apical plasma membrane of thyroid epithelial cells (Carvalho and Dupuy 2013; Rousset et al. 2015). In the male reproductive system, thyroid dysfunction is known to be associated with male infertility, although the functional mechanism of thyroid hormones remains unclear (Mazzilli et al. 2023). In fact, both iodine deficiency and excess can negatively affect semen quality and male fertility and prolong the time to pregnancy (Lisovskaya et al. 2021; Partal-Lorente et al. 2017; Sun et al. 2020). Given the relationship between iodine levels and thyroid and male reproductive functions (Farebrother et al. 2019; Lisovskaya et al. 2021; Partal-Lorente et al. 2017), dietary iodine intake, especially daily intake that may result in excessive iodine levels, should be carefully assessed and may require the development of better assessment strategies. Furthermore, whether iodine impacts male fertility through thyroid hormone is debatable, as elevated iodine levels have been reported in semen and in urine of infertile men, regardless of the level of their thyroid hormones (Partal-Lorente et al. 2017). The implication of EVs in thyroid physiology and pathophysiology as well as in thyroid dysfunction diagnosis and treatment has been discussed (Delcorte et al. 2022). However, research on the effect of iodine on male reproduction is scarce. More studies are required to better understand the role of iodine in male reproduction, especially regarding epididymal sperm maturation and the involvement of EVs.

4.8 Malnutrition, Hidden Hunger, and Male Infertility

Malnutrition is a condition resulting from deficient or excessive intake of nutrients, an imbalance of essential nutrients, or impaired nutrient utilization, as stated by the World Health Organization (WHO). Malnutrition includes undernutrition and overnutrition. The term undernutrition broadly encompasses four forms: stunting, wasting, underweight and deficiencies in essential vitamins and minerals; whereas the term overnutrition is generally characterized by overweight and obesity (Malnutrition Fact Sheets, WHO). According to the fact sheets of the WHO, nearly half of the deaths among children under 5 years old are linked to undernutrition, while in developed countries, overnutrition is common. In fact, overnutrition and undernutrition, so-called double burden of malnutrition, can coexist at all levels of the populations, regardless of origins of countries, cities, communities, or individuals (The Lancet 2020). This situation has been under-recognized and is a significant global health issue (Malnutrition Fact Sheets, WHO). Nutritional homeostatic regulation and decreased male fertility as well as epididymal function have been linked to some extent (Cooper et al. 1992; Liu et al. 2022; Ma et al. 2019; Mutalip 2019; Pruneda et al. 2007; Shum et al. 2022; Turner 2002; Vanderhout et al. 2021). The influence of dietary nutrition and the association of gut microbiota with male reproduction and infertility has also attracted attention in recent years (Almujaydil 2023; Argaw-Denboba et al. 2024; Cai et al. 2022; Panth et al. 2018; Pecora et al. 2023; Skoracka et al. 2020; Tully et al. 2024; Vanderhout et al. 2021). However, there are no reports on the functional correlation of double burden of malnutrition with male reproductive issues, and its relationship with male infertility, particularly idiopathic infertility, remains to be explored.

Hidden hunger may underlie male idiopathic infertility and unexplained epididymis dysfunction. As a form of malnutrition, hidden hunger, also known as micronutrient deficiency, refers to the lack of essential micronutrients in the diet, often without overt symptoms and can affect both undernourished and obese individuals (Das and Padhani 2022; Weffort and Lamounier 2024). Even without inadequate energy intake, hidden hunger can occur due to consuming a diet that is high in calories but poor in nutrients, especially vitamins and minerals in a balanced profile (Cheung 2015). Food security is commonly linked to hidden hunger (Lowe 2021). It has been reported that hidden hunger with deficiencies in iron, vitamin A, vitamin D, and zinc is prevalent, which can lead to serious health consequences, particularly in children (Weffort and Lamounier 2024). As mentioned, micronutrients have been specifically identified as important players in sperm maturation and epididymal function in a wide range of physiological processes. In addition, research indicates that certain forms of hinder hunger, even before perception, may negatively affect male fertility (Almujaydil 2023; Ferramosca and Zara

2022; Tully et al. 2024). For example, in an attempt to understand the mineral biology in epididymal sperm, our previous study used single-cell inductively-coupled plasma-mass spectrometry technique to analyze the elementomic profiles of epididymal spermatozoa from both wild-type and dysfunctional calcium-pump mice during capacitation (Zhang et al. 2023). In this study, we found that the macronutrient calcium was significantly changed as expected in impaired PMCA4-KO spermatozoa, and strikingly, the redox-labile elements including copper, iron, manganese, selenium, and zinc, were also significantly altered in dysfunctional PMCA4-deficient spermatozoa compared to wild-type spermatozoa (Zhang et al. 2023). Overall, the functional roles of micronutrients and the underlying cellular and molecular mechanisms of their deficiencies in epididymal function and sperm maturation warrant further investigations. Micronutrient interventions could also be used as an alternative for reproductive health. This approach may lead to new alternative strategies for addressing male infertility and reproductive health issues, especially idiopathic male infertility and unexplained epididymal dysfunction.

5 Conclusion Remarks

As one of the indispensable organs in male reproductive system, the epididymis is essentail for sperm maturation and the posttesticular reproductive processes. Regulating the balance of various essenital nutrients within the ideal and necessary range is crutial for epididymis function and sperm maturation, especially those redox-active micronutrients that are directly or indirectly involved in redox mebabolic processes. Functioning as the intercellualr carriers, epididymal EVs play important roles in epididymal sperm maturation, but research of epididymal EVs is still in its infancy. The epididymal luminal microenvironment has unique pro-reductive acidic and decalcitonic properties that are tailored for the redox metabolic demands of epididymal sperm maturation, allowing EVs to adapt to these processes. In this regard, there are still large gaps in our knowledge regarding the functional roles of epididymal EVs before their potenital use in diagnosis, therapy, and heath intervention. In the specific biological domain regarding the functional roles of micronutrients, i.e., vitamins and minerals relavant to the epididymal function and sperm maturation, a systemic and up-to-date review remains absent. Given the special microenvironment in epididymal lumen, i.e., acidic and decalcitonic features, we discussed how the proton availability may influence redox equilibria, in turn, redox metabolism in the epididymis, especially in reactions involving proton-coupled electron transfer, which is widely implicated in the metabolism of micronutrients. In addition, an increasing number of studies have shown that micronutrient deficiency and hinden hunger are related to reproductive health, whereas idiopathic male infertility still affects up to 15% of couples of reproductive age worldwide, so the redox metabolism of micronutrients deserves more attention and investigation. Understanding the vitaminology and mineralogy of the intriguiing intercellular interaction in the epididymis and resided spermatopzoa as well as their interplay network in redox metabolism can also help us to further understand reproductive biology. Thus, this chapter may provide some insights into these specific aspects. In conclusion, while EVs and micronutrients in epididymal function and redox metabolism represent a worthwhile area of research that could advance our knowledge, and EVs have high potenital applications in the development of thrapeutics and diagnostics, elucidating the precise mechanisms of micronutrients, whether or not EVs are invovled, warrants further investigation and understanding.

Declarations of Competing Interest None.

References

Abbaspour N, Hurrell R, Kelishadi R (2014) Review on iron and its importance for human health. J Res Med Sci 19:164–174
Agarwal A, Virk G, Ong C, du Plessis SS (2014) Effect of oxidative stress on male reproduction. World J Mens Health 32:1–17
Agarwal A, Baskaran S, Panner Selvam MK, Finelli R, Barbarosie C, Robert KA, Iovine C, Master K, Henkel R (2020) Scientific landscape of oxidative stress in male reproductive research: A scientometric study. Free Radic Biol Med 156:36–44
Agarwal A, Baskaran S, Parekh N, Cho CL, Henkel R, Vij S, Arafa M, Panner Selvam MK, Shah R (2021) Male infertility. Lancet 397:319–333
Agre P (2004) Aquaporin water channels (Nobel lecture). Angew Chem Int Ed Engl 43:4278–4290
Aitken R, Curry B (2011) Redox regulation of human sperm function: from the physiological control of sperm capacitation to the etiology of infertility and DNA damage in the germ line. Antioxid Redox Signal 14:367–381
Aitken RJ, Drevet JR (2020) The importance of oxidative stress in determining the functionality of mammalian spermatozoa: a two-edged sword. Antioxidants (Basel) 9:111
Aitken RJ, Vernet P (1998) Maturation of redox regulatory mechanisms in the epididymis. J Reprod Fertil Suppl 53:109–118
Aitken RJ, Smith TB, Jobling MS, Baker MA, De Iuliis GN (2014) Oxidative stress and male reproductive health. Asian J Androl 16:31–38
Akram S, Teong HF, Fliegel L, Pervaiz S, Clement MV (2006) Reactive oxygen species-mediated regulation of the Na+-H+ exchanger 1 gene expression connects intracellular redox status with cells' sensitivity to death triggers. Cell Death Differ 13:628–641
Ali W, Deng K, Bian Y, Liu Z, Zou H (2023) Spectacular role of epididymis and bio-active cargo of nano-scale exosome in sperm maturation: a review. Biomed Pharmacother 164:114889
Almujaydil MS (2023) The role of dietary nutrients in male infertility: a review. Life (Basel) 13:519
Alves MBR, Celeghini ECC, Belleannee C (2020) From sperm motility to sperm-borne microRNA signatures: new approaches to predict male fertility potential. Front Cell Dev Biol 8:791
Amann RP, Howards SS (1980) Daily spermatozoal production and epididymal spermatozoal reserves of the human male. J Urol 124:211–215
Ames BN (2018) Prolonging healthy aging: longevity vitamins and proteins. Proc Natl Acad Sci USA 115:10836–10844
Andriopoulos B Jr, Corradini E, Xia Y, Xia Y, Faasse SA, Faasse SA, Chen S, Chen S, Grgurevic L, Grgurevic L, Knutson MD, Knutson Md Pietrangelo A, Pietrangelo A, Vukicevic S, Vukicevic S, Lin HY, Babitt JL et al (2009) BMP6 is a key endogenous regulator of hepcidin expression and iron metabolism. Nat Genet 41:482
Angelova MI, Bitbol AF, Seigneuret M, Staneva G, Kodama A, Sakuma Y, Kawakatsu T, Imai M, Puff N (2018) pH sensing by lipids in membranes: the fundamentals of pH-driven migration,

polarization and deformations of lipid bilayer assemblies. Biochim Biophys Acta Biomembr 1860:2042–2063

Aparicio-Trejo OE, Aranda-Rivera AK, Osorio-Alonso H, Martinez-Klimova E, Sanchez-Lozada LG, Pedraza-Chaverri J, Tapia E (2022) Extracellular vesicles in redox signaling and metabolic regulation in chronic kidney disease. Antioxidants (Basel) 11:356

Argaw-Denboba A, Schmidt TSB, Di Giacomo M, Ranjan B, Devendran S, Mastrorilli E, Lloyd CT, Pugliese D, Paribeni V, Dabin J et al (2024) Paternal microbiome perturbations impact offspring fitness. Nature 629:652

Arredondo M, Nunez MT (2005) Iron and copper metabolism. Mol Asp Med 26:313–327

Aschner JL, Aschner M (2005) Nutritional aspects of manganese homeostasis. Mol Asp Med 26: 353–362

Asensi-Fabado MA, Munne-Bosch S (2010) Vitamins in plants: occurrence, biosynthesis and antioxidant function. Trends Plant Sci 15:582–592

Au CL, Wong PY (1980) Luminal acidification by the perfused rat cauda epididymidis. J Physiol 309:419–427

Augiere C, Campolina-Silva G, Vijayakumaran A, Medagedara O, Lavoie-Ouellet C, Joly Beauparlant C, Droit A, Barrachina F, Ottino K, Battistone MA et al (2024) ARL13B controls male reproductive tract physiology through primary and Motile Cilia. Commun Biol 7:1318

Avila DS, Puntel RL, Aschner M (2013) Manganese in health and disease. Met Ions Life Sci 13: 199–227

Ayala A, Munoz MF, Arguelles S (2014) Lipid peroxidation: production, metabolism, and signaling mechanisms of malondialdehyde and 4-hydroxy-2-nonenal. Oxidative Med Cell Longev 2014:360438

Bai X, Wang P (2022) Relationship between sperm NAD + concentration and reproductive aging in normozoospermia men: A Cohort study. BMC Urol 22:159

Baj J, Flieger W, Barbachowska A, Kowalska B, Flieger M, Forma A, Teresinski G, Portincasa P, Buszewicz G, Radzikowska-Buchner E et al (2023) Consequences of disturbing manganese homeostasis. Int J Mol Sci 24:14959

Banihani SA (2017) Vitamin B(12) and semen quality. Biomolecules 7:42

Banihani SA, Aljabali SM (2020) Seminal plasma vitamin B(6) levels in men with asthenozoospermia and men with normal sperm motility, a measurement using liquid chromatography with tandem mass spectrometry. Andrologia 52:e13556

Barrachina F, Battistone MA, Castillo J, Mallofre C, Jodar M, Breton S, Oliva R (2022) Sperm acquire epididymis-derived proteins through epididymosomes. Hum Reprod 37:651–668

Baskaran S, Finelli R, Agarwal A, Henkel R (2020a) Reactive oxygen species in male reproduction: a boon or a bane? Andrologia 53:e13577

Baskaran S, Panner Selvam MK, Agarwal A (2020b) Exosomes of male reproduction. Adv Clin Chem 95:149–163

Battistone MA, Merkulova M, Park YJ, Peralta MA, Gombar F, Brown D, Breton S (2019) Unravelling purinergic regulation in the epididymis: activation of V-ATPase-dependent acidification by luminal ATP and adenosine. J Physiol 597:1957–1973

Battistone MA, Elizagaray ML, Barrachina F, Ottino K, Mendelsohn AC, Breton S (2024) Immunoregulatory mechanisms between epithelial clear cells and mononuclear phagocytes in the epididymis. Andrology-Us 12:949–963

Belardin LB, Brochu K, Legare C, Breton S (2024) Regulation of NHE3 subcellular localization in epididymal principal cells: pH, cyclic adenosine 3,5 monophosphate (cAMP), and adenosine signaling. Andrology. https://doi.org/10.1111/andr.13820

Belladelli F, Muncey W, Seranio N, Eisenberg ML (2023) Counseling for the man with severe male infertility. Curr Opin Urol 33:5–9

Belleannee C (2015) Extracellular microRNAs from the epididymis as potential mediators of cell-to-cell communication. Asian J Androl 17:730–736

Belleannee C, Belghazi M, Labas V, Teixeira-Gomes AP, Gatti JL, Dacheux JL, Dacheux F (2011) Purification and identification of sperm surface proteins and changes during epididymal maturation. Proteomics 11:1952–1964
Belleannee C, Calvo E, Caballero J, Sullivan R (2013) Epididymosomes convey different repertoires of microRNAs throughout the bovine epididymis. Biol Reprod 89:30
Beltrame FL, de Santi F, Vendramini V, Cabral REL, Miraglia SM, Cerri PS, Sasso-Cerri E (2019) Vitamin B(12) prevents cimetidine-induced androgenic failure and damage to sperm quality in rats. Front Endocrinol (Lausanne) 10:309
Beltran FA, Torres-Diaz LL, Troncoso-Escudero P, Villalobos-Gonzalez J, Mayorga-Weber G, Lara M, Covarrubias-Pinto A, Valdivia S, Vicencio I, Papic E et al (2025) Distinct roles of ascorbic acid in extracellular vesicles and free form: implications for metabolism and oxidative stress in presymptomatic Huntington's disease. Free Radic Biol Med 227:521–535
Bernet A, Bastien A, Soulet D, Jerczynski O, Roy C, Bianchi Rodrigues Alves M, Lecours C, Tremblay ME, Bailey JL, Robert C et al (2018) Cell-lineage specificity of primary cilia during postnatal epididymal development. Hum Reprod 33:1829–1838
Bhat OM, Yuan X, Li G, Lee R, Li PL (2018) Sphingolipids and redox signaling in renal regulation and chronic kidney diseases. Antioxid Redox Signal 28:1008–1026
Bhattarai KR, Riaz TA, Kim HR, Chae HJ (2021) The aftermath of the interplay between the endoplasmic reticulum stress response and redox signaling. Exp Mol Med 53:151–167
Bienert GP, Chaumont F (2014) Aquaporin-facilitated transmembrane diffusion of hydrogen peroxide. Biochim Biophys Acta 1840:1596–1604
Bisserier M, Brojakowska A, Saffran N, Rai AK, Lee B, Coleman M, Sebastian A, Evans A, Mills PJ, Addya S et al (2022) Astronauts plasma-derived exosomes induced aberrant EZH2-mediated H3K27me3 epigenetic regulation of the vitamin D receptor. Front Cardiovasc Med 9:855181
Bister N, Pistono C, Huremagic B, Jolkkonen J, Giugno R, Malm T (2020) Hypoxia and extracellular vesicles: A review on methods, vesicular cargo and functions. J Extracell Vesicles 10: e12002
Bjorkgren I, Gylling H, Turunen H, Huhtaniemi I, Strauss L, Poutanen M, Sipila P (2015) Imbalanced lipid homeostasis in the conditional Dicer1 knockout mouse epididymis causes instability of the sperm membrane. FASEB J 29:433–442
Blomberg Jensen M, Nielsen JE, Jorgensen A, Rajpert-De Meyts E, Kristensen DM, Jorgensen N, Skakkebaek NE, Juul A, Leffers H (2010) Vitamin D receptor and vitamin D metabolizing enzymes are expressed in the human male reproductive tract. Hum Reprod 25:1303–1311
Blomberg Jensen M, Bjerrum PJ, Jessen TE, Nielsen JE, Joensen UN, Olesen IA, Petersen JH, Juul A, Dissing S, Jorgensen N (2011) Vitamin D is positively associated with sperm motility and increases intracellular calcium in human spermatozoa. Hum Reprod 26:1307–1317
Blomqvist SR, Vidarsson H, Soder O, Enerback S (2006) Epididymal expression of the forkhead transcription factor Foxi1 is required for male fertility. EMBO J 25:4131–4141
Boilard E (2018) Extracellular vesicles and their content in bioactive lipid mediators: more than a sack of microRNA. J Lipid Res 59:2037–2046
Borodina I, Kenny LC, McCarthy CM, Paramasivan K, Pretorius E, Roberts TJ, van der Hoek SA, Kell DB (2020) The biology of ergothioneine, an antioxidant nutraceutical. Nutr Res Rev 33: 190–217
Bosworth CR, Levin G, Robinson-Cohen C, Hoofnagle AN, Ruzinski J, Young B, Schwartz SM, Himmelfarb J, Kestenbaum B, de Boer IH (2012) The serum 24,25-dihydroxyvitamin D concentration, a marker of vitamin D catabolism, is reduced in chronic kidney disease. Kidney Int 82:693–700
Bouillon R, Schuit F, Antonio L, Rastinejad F (2019) Vitamin D binding protein: A historic overview. Front Endocrinol (Lausanne) 10:910
Brada N, Gordon MM, Shao JS, Wen J, Alpers DH (2000) Production of gastric intrinsic factor, transcobalamin, and haptocorrin in opossum kidney cells. Am J Physiol Renal Physiol 279: F1006–F1013

Braun RE (1998) Post-transcriptional control of gene expression during spermatogenesis. Semin Cell Dev Biol 9:483–489

Breton S, Brown D (2013) Regulation of luminal acidification by the V-ATPase. Physiology (Bethesda) 28:318–329

Breton S, Smith PJ, Lui B, Brown D (1996) Acidification of the male reproductive tract by a proton pumping (H+)-ATPase. Nat Med 2:470–472

Breton S, Nair AV, Battistone MA (2019) Epithelial dynamics in the epididymis: role in the maturation, protection, and storage of spermatozoa. Andrology 7:631–643

Briffa J, Sinagra E, Blundell R (2020) Heavy metal pollution in the environment and their toxicological effects on humans. Heliyon 6:e04691

Brown D, Breton S (1996) Mitochondria-rich, proton-secreting epithelial cells. J Exp Biol 199: 2345–2358

Cabrillana ME, Monclus MLA, Lancellotti TES, Boarelli PV, Vincenti AE, Fornes MM, Sanabria EA, Fornes MW (2017) Thiols of flagellar proteins are essential for progressive motility in human spermatozoa. Reprod Fertil Dev 29:1435–1446

Cai H, Cao X, Qin D, Liu Y, Liu Y, Hua J, Peng S (2022) Gut microbiota supports male reproduction via nutrition, immunity, and signaling. Front Microbiol 13:977574

Camacho-Pereira J, Tarrago MG, Chini CCS, Nin V, Escande C, Warner GM, Puranik AS, Schoon RA, Reid JM, Galina A et al (2016) CD38 dictates age-related NAD decline and mitochondrial dysfunction through an SIRT3-dependent mechanism. Cell Metab 23:1127–1139

Can B, Kulaksiz Erkmen G, Dalmizrak O, Ogus IH, Ozer N (2010) Purification and characterisation of rat kidney glutathione reductase. Protein J 29:250–256

Carmel R, Bernstein GS (1984) Transcobalamin II in human seminal plasma. J Clin Invest 73:868–872

Carr A, Frei B (1999) Does vitamin C act as a pro-oxidant under physiological conditions? FASEB J 13:1007–1024

Carr DW, Usselman MC, Acott TS (1985) Effects of pH, lactate, and viscoelastic drag on sperm motility: a species comparison. Biol Reprod 33:588–595

Carter DE (1995) Oxidation-reduction reactions of metal ions. Environ Health Perspect 103(Suppl 1):17–19

Carvalho DP, Dupuy C (2013) Role of the NADPH oxidases DUOX and NOX4 in thyroid oxidative stress. Eur Thyroid J 2:160–167

Catalano M, O'Driscoll L (2020) Inhibiting extracellular vesicles formation and release: a review of EV inhibitors. J Extracell Vesicles 9:1703244

Chan JC, Morgan CP, Adrian Leu N, Shetty A, Cisse YM, Nugent BM, Morrison KE, Jasarevic E, Huang W, Kanyuch N et al (2020) Reproductive tract extracellular vesicles are sufficient to transmit intergenerational stress and program neurodevelopment. Nat Commun 11:1499

Chatterjee V, Yang X, Ma Y, Wu MH, Yuan SY (2020) Extracellular vesicles: new players in regulating vascular barrier function. Am J Physiol Heart Circ Physiol 319:H1181–H1196

Chen H, Zhao HX, Huang XF, Chen GW, Yang ZX, Sun WJ, Tao MH, Yuan Y, Wu JQ, Sun F et al (2012) Does high load of oxidants in human semen contribute to male factor infertility? Antioxid Redox Signal 16:754–759

Chen Q, Yan M, Cao Z, Li X, Zhang Y, Shi J, Feng GH, Peng H, Zhang X, Zhang Y et al (2016a) Sperm tsRNAs contribute to intergenerational inheritance of an acquired metabolic disorder. Science 351:397–400

Chen Q, Yan W, Duan EK (2016b) Epigenetic inheritance of acquired traits through sperm RNAs and sperm RNA modifications. Nat Rev Genet 17:733–743

Chen P, Bornhorst J, Aschner M (2018) Manganese metabolism in humans. Front Biosci (Landmark Ed) 23(9):1655

Chen X, Zhou Y, Yu J (2019) Exosome-like nanoparticles from ginger rhizomes inhibited NLRP3 Inflammasome activation. Mol Pharm 16:2690–2699

Chen H, Alves MBR, Belleannee C (2021) Contribution of epididymal epithelial cell functions to sperm epigenetic changes and the health of progeny. Hum Reprod Update 28:51–66

Chen L, Min J, Wang F (2022) Copper homeostasis and cuproptosis in health and disease. Signal Transduct Target Ther 7:378
Chen Q, Zhou R, Yang C, Jiang Q, Yuan H, Qiu X, Tian H, Zhou J, Liu C (2023) Ergothioneine attenuates varicocele-induced testicular damage by upregulating HSP90AA1 in rats. J Biochem Mol Toxicol 37:e23301
Chen B, Yu P, Chan WN, Xie F, Zhang Y, Liang L, Leung KT, Lo KW, Yu J, Tse GMK et al (2024) Cellular zinc metabolism and zinc signaling: from biological functions to diseases and therapeutic targets. Signal Transduct Target Ther 9:6
Cheung KH, Leung GP, Leung MC, Shum WW, Zhou WL, Wong PY (2005) Cell-cell interaction underlies formation of fluid in the male reproductive tract of the rat. J Gen Physiol 125:443–454
Cheung HC, Shen A, Oo S, Tilahun H, Cohen MJ, Berkowitz SA (2015) Food insecurity and body mass index: a longitudinal mixed methods study, Chelsea, Massachusetts, 2009–2013. Prev Chronic Dis 12:E125. https://doi.org/10.5888/pcd12.150001. PMID: 26247425; PMCID: PMC4565511
Chew SB, Leung GP, Leung PY, Tse CM, Wong PY (2000) Polarized distribution of NHE1 and NHE2 in the rat epididymis. Biol Reprod 62:755–758
Chiaradia E, Tancini B, Emiliani C, Delo F, Pellegrino RM, Tognoloni A, Urbanelli L, Buratta S (2021) Extracellular vesicles under oxidative stress conditions: biological properties and physiological roles. Cells 10:1763
Chrétien DW, Benit P, Ha HH, Keipert S, El-Khoury R, Chang YT, Jastroch M, Jacobs HT, Rustin P, Rak M (2018) Mitochondria are physiologically maintained at close to 50 degrees C. PLoS Biol 16:e2003992
Chu C, Zhang YL, Yu L, Sharma S, Fei ZL, Drevet JR (2019) Epididymal small non-coding RNA studies: progress over the past decade. Andrology 7:681–689
Chu X, Liu D, Li T, Ke H, Xin D, Wang S, Cao Y, Xue H, Wang Z (2020) Hydrogen sulfide-modified extracellular vesicles from mesenchymal stem cells for treatment of hypoxic-ischemic brain injury. J Control Release 328:13–27
Chubanava S, Karavaeva I, Ehrlich AM, Justicia RM, Basse AL, Kulik I, Dalbram E, Ahwazi D, Heaselgrave SR, Trošt K et al (2025) NAD depletion in skeletal muscle does not compromise muscle function or accelerate aging. Cell Metab 37:1460
Cirino G, Szabo C, Papapetropoulos A (2023) Physiological roles of hydrogen sulfide in mammalian cells, tissues, and organs. Physiol Rev 103:31–276
Clarke-Bland CE, Bill RM, Devitt A (2022) Emerging roles for AQP in mammalian extracellular vesicles. Biochim Biophys Acta Biomembr 1864:183826
Colagar AH, Marzony ET (2009) Ascorbic acid in human seminal plasma: determination and its relationship to sperm quality. J Clin Biochem Nutr 45:144–149
Conine CC, Sun F, Song L, Rivera-Perez JA, Rando OJ (2018) Small RNAs gained during Epididymal transit of sperm are essential for embryonic development in mice. Dev Cell 46(470–480):e473
Conte C, Cataldi S, Arcuri C, Mirarchi A, Lazzarini A, Garcia-Gil M, Beccari T, Curcio F, Albi E (2021) Vitamin D3 enriches ceramide content in exosomes released by embryonic hippocampal cells. Int J Mol Sci 22:9287
Cooper TG (2005) Cytoplasmic droplets: the good, the bad or just confusing? Hum Reprod 20:9–11
Cooper TG (2011) The epididymis, cytoplasmic droplets and male fertility. Asian J Androl 13:130–138
Cooper TG, Raczek S, Yeung CH, Schwab E, Schulze H, Hertle L (1992) Composition of fluids obtained from human epididymal cysts. Urol Res 20:275–280
Corinti D, Chiavarino B, Scuderi D, Fraschetti C, Filippi A, Fornarini S, Crestoni ME (2021) Molecular properties of bare and microhydrated vitamin B5-calcium complexes. Int J Mol Sci 22:692
Cornejo-Guerra C, Salazar-Ardiles C, Morales P, Andrade DC (2024) Consequences of exposure to hypobaric hypoxia associated with high altitude on spermatogenesis and seminal parameters: A literature review. Cells 13:592

Cornwall GA (2009) New insights into epididymal biology and function. Hum Reprod Update 15: 213–227

Cornwall GA, Do HQ, Hewetson A, Muthusubramanian A, Myers C (2019) The epididymal amyloid matrix: structure and putative functions. Andrology 7:603–609

Corsini CA-O, Boeri LA-OX, Candela LA-O, Pozzi EA-O, Belladelli FA-OX, Capogrosso PA-O, Fallara GA-O, Schifano NA-O, Cignoli DA-O, Ventimiglia EA-O et al (2022) Is there a relevant clinical impact in differentiating idiopathic versus unexplained male infertility?. World J Mens Health, vol 41, p 354

Covarrubias AJ, Perrone R, Grozio A, Verdin E (2021) NAD(+) metabolism and its roles in cellular processes during ageing. Nat Rev Mol Cell Biol 22:119–141

Crha I, Kralikova M, Melounova J, Ventruba P, Zakova J, Beharka R, Husicka R, Pohanka M, Huser M (2010) Seminal plasma homocysteine, folate and cobalamin in men with obstructive and non-obstructive azoospermia. J Assist Reprod Genet 27:533–538

Cui T, Terlecki R (2018) Prevalence of relative deficiencies in testosterone and vitamin B12 among patients referred for chronic Orchialgia: implications for management. Am J Mens Health 12(3): 608–611

Cui C, Jiang M, Jain N, Das S, Lo YH, Kermani AA, Pipatpolkai T, Sun J (2024) Structural basis of human NOX5 activation. Nat Commun 15:3994

Culotta V (2010) Cell biology of copper. JBIC J Biol Inorg Chem 15:1–2

Da Silva N, Pietrement C, Brown D, Breton S (2006) Segmental and cellular expression of aquaporins in the male excurrent duct. BBA-Biomembranes 1758:1025–1033

Da Silva N, Shum WW, El-Annan J, Paunescu TG, McKee M, Smith PJ, Brown D, Breton S (2007) Relocalization of the V-ATPase B2 subunit to the apical membrane of epididymal clear cells of mice deficient in the B1 subunit. Am J Physiol Cell Physiol 293:C199–C210

Da Silva N, Cortez-Retamozo V, Reinecker HC, Wildgruber M, Hill E, Brown D, Swirski FK, Pittet MJ, Breton S (2011) A dense network of dendritic cells populates the murine epididymis. Reproduction 141:653–663

Dacheux JL, Dacheux F (2014) New insights into epididymal function in relation to sperm maturation. Reproduction 147:R27–R42

Darr CR, Varner DD, Teague S, Cortopassi GA, Datta S, Meyers SA (2016) Lactate and pyruvate are major sources of energy for stallion sperm with dose effects on mitochondrial function, motility, and ROS production. Biol Reprod 95:34

Das JK, Padhani ZA (2022) Alleviating hidden hunger: an infallible bridge to improved health and nutrition. Lancet Glob Health 10:e1539–e1540

Das P, Choudhari AR, Dhawan A, Singh R (2009) Role of ascorbic acid in human seminal plasma against the oxidative damage to the sperms. Indian J Clin Biochem 24:312–315

Das S, Extracellular RNACC, Ansel KM, Bitzer M, Breakefield XO, Charest A, Galas DJ, Gerstein MB, Gupta M, Milosavljevic A et al (2019) The Extracellular RNA communication consortium: establishing foundational knowledge and Technologies for Extracellular RNA research. Cell 177:231–242

Delcorte O, Degosserie J, Pierreux CE (2022) Role of Extracellular vesicles in thyroid physiology and diseases: implications for diagnosis and treatment. Biomedicine 10:2585

Dereven'kov IA, Salnikov DS, Makarov SV, Boss GR, Koifman OI (2013) Kinetics and mechanism of oxidation of super-reduced cobalamin and cobinamide species by thiosulfate, sulfite and dithionite. Dalton Trans 42:15307–15316

Dhakal J, Sharma CS, Nannapaneni R, Mc DC, Kim T, Kiess A (2019) Effect of chlorine-induced sublethal oxidative stress on the biofilm-forming ability of salmonella at different temperatures, nutrient conditions, and substrates. J Food Prot 82:78–92

Diaz M, Mesa-Herrera F, Marin R (2021) DHA and its elaborated modulation of antioxidant defenses of the brain: implications in aging and AD neurodegeneration. Antioxidants (Basel) 10:10

Djakiew D, Cardullo R (1986) Lower temperature of the cauda epididymidis facilitates the storage of sperm by enhancing oxygen availability. Gamete Res 15:237–245

Durairajanayagam D, Singh D, Agarwal A, Henkel R (2021) Causes and consequences of sperm mitochondrial dysfunction. Andrologia 53:e13666
Dutta S, Majzoub A, Agarwal A (2019) Oxidative stress and sperm function: A systematic review on evaluation and management. Arab J Urol 17:87–97
Elfgen V, Mietens A, Mewe M, Hau T, Middendorff R (2018) Contractility of the epididymal duct: function, regulation and potential drug effects. Reproduction 156:R125–R141
Esworthy RS, Doroshow JH, Chu FF (2022) The beginning of GPX2 and 30 years later. Free Radic Biol Med 188:419–433
Fang J, Liang W (2021) ASCs -derived exosomes loaded with vitamin A and quercetin inhibit rapid senescence-like response after acute liver injury. Biochem Biophys Res Commun 572:125–130
Farebrother J, Zimmermann MB, Andersson M (2019) Excess iodine intake: sources, assessment, and effects on thyroid function. Ann N Y Acad Sci 1446:44–65
Fazzio A, Caponnetto A, Ferrara C, Purrello M, Di Pietro C, Battaglia R (2024) From germ cells to implantation: the role of Extracellular vesicles. J Dev Biol 12:22
Federici C, Shahaj E, Cecchetti S, Camerini S, Casella M, Iessi E, Camisaschi C, Paolino G, Calvieri S, Ferro S et al (2020) Natural-killer-derived extracellular vesicles: immune sensors and interactors. Front Immunol 11:262
Feelisch M, Cortese-Krott MM, Santolini J, Wootton SA, Jackson AA (2022) Systems redox biology in health and disease. EXCLI J 21:623–646
Feng YQ, Liu X, Zuo N, Yu MB, Bian WM, Han BQ, Sun ZY, De Felici M, Shen W, Li L (2024) NAD(+) precursors promote the restoration of spermatogenesis in busulfan-treated mice through inhibiting Sirt2-regulated ferroptosis. Theranostics 14:2622–2636
Ferramosca A, Zara V (2022) Diet and male fertility: the impact of nutrients and antioxidants on sperm energetic metabolism. Int J Mol Sci 23:2542
Ferrell M, Wang Z, Anderson JT, Li XS, Witkowski M, DiDonato JA, Hilser JR, Hartiala JA, Haghikia A, Cajka T et al (2024) A terminal metabolite of niacin promotes vascular inflammation and contributes to cardiovascular disease risk. Nat Med 30:424–434
Fleming JC, Tartaglini E, Kawatsuji R, Yao D, Fujiwara Y, Bednarski JJ, Fleming MD, Neufeld EJ (2003) Male infertility and thiamine-dependent erythroid hypoplasia in mice lacking thiamine transporter Slc19a2. Mol Genet Metab 80:234–241
Foot NJ, Gonzalez MB, Gembus K, Fonseka P, Sandow JJ, Nguyen TT, Tran D, Webb AI, Mathivanan S, Robker RL et al (2021) Arrdc4-dependent extracellular vesicle biogenesis is required for sperm maturation. J Extracell Vesicles 10:e12113
Fowler SL, Behr TS, Turkes E, O'Brien DP, Cauhy PM, Rawlinson I, Edmonds M, Foiani MS, Schaler A, Crowley G et al (2024) Tau filaments are tethered within brain extracellular vesicles in Alzheimer's disease. Nat Neurosci 28:40–48
Fraczek M, Kurpisz M (2015) Mechanisms of the harmful effects of bacterial semen infection on ejaculated human spermatozoa: potential inflammatory markers in semen. Folia Histochem Cytobiol 53:201–217
Franco CN, Seabrook LJ, Nguyen ST, Leonard JT, Albrecht LV (2022) Simplifying the B complex: how vitamins B6 and B9 modulate one carbon metabolism in cancer and beyond. Metabolites 12:961
Frenette G, Sullivan R (2001) Prostasome-like particles are involved in the transfer of P25b from the bovine epididymal fluid to the sperm surface. Mol Reprod Dev 59:115–121
Frenette G, Lessard C, Sullivan R (2002) Selected proteins of "prostasome-like particles" from epididymal cauda fluid are transferred to epididymal caput spermatozoa in bull. Biol Reprod 67: 308–313
Frenette G, Lessard C, Madore E, Fortier MA, Sullivan R (2003) Aldose reductase and macrophage migration inhibitory factor are associated with epididymosomes and spermatozoa in the bovine epididymis. Biol Reprod 69:1586–1592
Frenette G, Girouard J, D'Amours O, Allard N, Tessier L, Sullivan R (2010) Characterization of two distinct populations of epididymosomes collected in the intraluminal compartment of the bovine cauda epididymis. Biol Reprod 83:473–480

Frizzell RA, Hanrahan JW (2012) Physiology of epithelial chloride and fluid secretion. Cold Spring Harb Perspect Med 2:a009563
Furie B, Bouchard BA, Furie BC (1999) Vitamin K-dependent biosynthesis of gamma-carboxyglutamic acid. Blood 93:1798–1808
Gai C, Li T, Zhao Y, Cheng Y, Song Y, Luo Q, Liu D, Wang Z (2024) Mesenchymal stromal cells deliver H(2)S-enhanced Nrf2 via extracellular vesicles to mediate mitochondrial homeostasis for repairing hypoxia-ischemia brain damage. Free Radic Biol Med 225:528–545
Galimov SN, Gromenko JY, Bulygin KV, Galimov KS, Galimova EF, Sinelnikov MY (2021) The level of secondary messengers and the redox state of NAD(+)/NADH are associated with sperm quality in infertility. J Reprod Immunol 148:103383
Gangolf M, Czerniecki J, Radermecker M, Detry O, Nisolle M, Jouan C, Martin D, Chantraine F, Lakaye B, Wins P et al (2010) Thiamine status in humans and content of phosphorylated thiamine derivatives in biopsies and cultured cells. PLoS One 5:e13616
Ganz T, Nemeth E (2012) Hepcidin and iron homeostasis. Biochim Biophys Acta 1823:1434–1443
Gao DY, Zhang BL, Leung MCT, Au SCL, Wong PYD, Shum WWC (2016) Coupling of TRPV6 and TMEM16A in epithelial principal cells of the rat epididymis. J Gen Physiol 148:161–182
Gao DD, Xu JW, Qin WB, Peng L, Qiu ZE, Wang LL, Lan CF, Cao XN, Xu JB, Zhu YX et al (2018) Cellular mechanism underlying hydrogen sulfide mediated epithelial K(+) secretion in rat epididymis. Front Physiol 9:1886
Garratt M, Bathgate R, de Graaf SP, Brooks RC (2013) Copper-zinc superoxide dismutase deficiency impairs sperm motility and in vivo fertility. Reproduction 146:297–304
Gaspar RS, Ferreira PM, Mitchell JL, Pula G, Gibbins JM (2021) Platelet-derived extracellular vesicles express NADPH oxidase-1 (Nox-1), generate superoxide and modulate platelet function. Free Radic Biol Med 165:395–400
Gastaldi G, Cova E, Verri A, Laforenza U, Faelli A, Rindi G (2000) Transport of thiamin in rat renal brush border membrane vesicles. Kidney Int 57:2043–2054
Gatti JL, Castella S, Dacheux F, Ecroyd H, Metayer S, Thimon V, Dacheux JL (2004) Post-testicular sperm environment and fertility. Anim Reprod Sci 82-83:321–339
Gatti JL, Metayer S, Belghazi M, Dacheux F, Dacheux JL (2005) Identification, proteomic profiling, and origin of ram epididymal fluid exosome-like vesicles. Biol Reprod 72:1452–1465
Gervasi MG, Visconti PE (2017) Molecular changes and signaling events occurring in spermatozoa during epididymal maturation. Andrology 5:204–218
Gibb Z, Griffin RA, Aitken RJ, De Iuliis GN (2020) Functions and effects of reactive oxygen species in male fertility. Anim Reprod Sci 220:106456
Gibb Z, Aitken RJ, Sheridan AR, Holt B, Waugh S, Swegen A (2024) The effects of oxidative stress and intracellular calcium on mitochondrial permeability transition pore formation in equine spermatozoa. FASEB BioAdvances 6:143
Girardet L, Augiere C, Asselin MP, Belleannee C (2019) Primary cilia: biosensors of the male reproductive tract. Andrology 7:588–602
Girouard J, Frenette G, Sullivan R (2009) Compartmentalization of proteins in epididymosomes coordinates the association of epididymal proteins with the different functional structures of bovine spermatozoa. Biol Reprod 80:965–972
Girouard J, Frenette G, Sullivan R (2011) Comparative proteome and lipid profiles of bovine epididymosomes collected in the intraluminal compartment of the caput and cauda epididymidis. Int J Androl 34:e475–e486
Gonzalez SN, Sulzyk V, Weigel Munoz M, Cuasnicu PS (2021) Cysteine-rich secretory proteins (CRISP) are key players in mammalian fertilization and fertility. Front Cell Dev Biol 9:800351
González-Chávez SA, Arévalo-Gallegos S, Rascón-Cruz Q (2009) Lactoferrin: structure, function and applications. Int J Antimicrob Agents 33:301.e301–301.e308
Gorlach A, Bertram K, Hudecova S, Krizanova O (2015) Calcium and ROS: A mutual interplay. Redox Biol 6:260–271

Goss DM, Vasilescu SA, Sacks G, Gardner DK, Warkiani ME (2023) Microfluidics facilitating the use of small extracellular vesicles in innovative approaches to male infertility. Nat Rev Urol 20: 66–95
Gr AT, Bhavani S, Kuppuraj S, Sen S, Udaya P, Babu Kannan N, Ramasamy K, Dharmalingam K, Banerjee D (2025) Retinol-binding protein 3 in vitreous and plasma-derived small extracellular vesicles is reduced in proliferative diabetic retinopathy. Exp Eye Res 255:110385
Greco E, Iacobelli M, Rienzi L, Ubaldi F, Ferrero S, Tesarik J (2005) Reduction of the incidence of sperm DNA fragmentation by oral antioxidant treatment. J Androl 26:349–353
Griffiths GS, Galileo DS, Reese K, Martin-Deleon PA (2008) Investigating the role of murine epididymosomes and uterosomes in GPI-linked protein transfer to sperm using SPAM1 as a model. Mol Reprod Dev 75:1627–1636
Grimalt P, Bertini F, Fornes MW (2000) High-affinity sites for beta-D-galactosidase on membrane-bound vesicles isolated from rat epididymal fluid. Arch Androl 44:85–91
Guthrie HD, Welch GR, Theisen DD, Woods LC 3rd (2011) Effects of hypothermic storage on intracellular calcium, reactive oxygen species formation, mitochondrial function, motility, and plasma membrane integrity in striped bass (Morone saxatilis) sperm. Theriogenology 75:951–961
Hadrup N, Ravn-Haren G (2023) Toxicity of repeated oral intake of organic selenium, inorganic selenium, and selenium nanoparticles: a review. J Trace Elem Med Biol 79:127235
Hamamah S, Gatti JL (1998) Role of the ionic environment and internal pH on sperm activity. Hum Reprod 13(Suppl 4):20–30
Han R, Zhou D, Ji N, Yin Z, Wang J, Zhang Q, Zhang H, Liu J, Liu X, Liu H et al (2025) Folic acid-modified ginger-derived extracellular vesicles for targeted treatment of rheumatoid arthritis by remodeling immune microenvironment via the PI3K-AKT pathway. J Nanobiotechnol 23:41
Hawkes WC, Turek PJ (2001) Effects of dietary selenium on sperm motility in healthy men. J Androl 22:764–772
Henkel R, Sandhu IS, Agarwal A (2019) The excessive use of antioxidant therapy: A possible cause of male infertility? Andrologia 51:e13162
Henzler T, Steudle E (2000) Transport and metabolic degradation of hydrogen peroxide in Chara corallina: model calculations and measurements with the pressure probe suggest transport of H (2)O(2) across water channels. J Exp Bot 51:2053–2066
Hermo L, Jacks D (2002) Nature's ingenuity: bypassing the classical secretory route via apocrine secretion. Mol Reprod Dev 63:394–410
Hermo L, Robaire B (2002) The epididymis: from molecules to clinical practice. Kluwer Academic/ Plenum Publishers, New York
Hermo L, Oliveira RL, Smith CE, Au CE, Bergeron JJM (2019) Dark side of the epididymis: tails of sperm maturation. Andrology 7:566–580
Hermo L, Oliveira R, Dufresne J, Gregory M, Cyr DG (2025) Basal and immune cells of the epididymis: an electron microscopy view of their association. Adv Exp Med Biol 1469:67–87
Hidalgo C, Donoso P (2008) Crosstalk between calcium and redox signaling: from molecular mechanisms to health implications. Antioxid Redox Signal 10:1275–1312
Hille R, Hall J, Basu P (2014) The mononuclear molybdenum enzymes. Chem Rev 114:3963–4038
Hinton BT (1990) The testicular and Epididymal luminal amino-acid microenvironment in the rat. J Androl 11:498–505
Hoek J, Steegers-Theunissen RPM, Willemsen SP, Schoenmakers S (2020) Paternal folate status and sperm quality, pregnancy outcomes, and epigenetics: a systematic review and meta-analysis. Mol Nutr Food Res 64:e1900696
Holick MF (2009) Vitamin D status: measurement, interpretation, and clinical application. Ann Epidemiol 19:73–78
Hondal RJ (2023) Selenium vitaminology: the connection between selenium, vitamin C, vitamin E, and ergothioneine. Curr Opin Chem Biol 75:102328
Hong CY, Chiang BN, Turner P (1984) Calcium-ion is the key regulator of human-sperm function. Lancet 2:1449–1451

Hosseinabadi F, Jenabi M, Ghafarizadeh AA, Yazdanikhah S (2020) The effect of vitamin B12 supplement on post-thaw motility, viability and DNA damage of human sperm. Andrologia 52: e13877

Hrubsa M, Siatka T, Nejmanova I, Voprsalova M, Kujovska Krcmova L, Matousova K, Javorska L, Macakova K, Mercolini L, Remiao F et al (2022) Biological properties of vitamins of the B-complex, Part 1: Vitamins B(1), B(2), B(3), and B(5). Nutrients 14:484

Hu ML, Chen YK, Lin YF (1995) The antioxidant and prooxidant activity of some B vitamins and vitamin-like compounds. Chem Biol Interact 97:63–73

Hu Q, Zhang Y, Lou H, Ou Z, Liu J, Duan W, Wang H, Ge Y, Min J, Wang F et al (2021) GPX4 and vitamin E cooperatively protect hematopoietic stem and progenitor cells from lipid peroxidation and ferroptosis. Cell Death Dis 12:706

Huang HM, Chen HL, Gibson GE (2010) Thiamine and oxidants interact to modify cellular calcium stores. Neurochem Res 35:2107–2116

Huszar G, Sbracia M, Vigue L, Miller DJ, Shur BD (1997) Sperm plasma membrane remodeling during spermiogenetic maturation in men: relationship among plasma membrane beta 1,4-galactosyltransferase, cytoplasmic creatine phosphokinase, and creatine phosphokinase isoform ratios. Biol Reprod 56:1020–1024

Ijiri TW, Vadnais ML, Huang AP, Lin AM, Levin LR, Buck J, Gerton GL (2013) Thiol changes during epididymal maturation: a link to flagellar angulation in mouse spermatozoa? Andrology 2:65–75

Ikawa M, Inoue N, Benham AM, Okabe M (2010) Fertilization: a sperm's journey to and interaction with the oocyte. J Clin Invest 120:984–994

Ingold I, Berndt C, Schmitt S, Doll S, Poschmann G, Buday K, Roveri A, Peng X, Porto Freitas F, Seibt T et al (2018) Selenium utilization by GPX4 is required to prevent Hydroperoxide-induced Ferroptosis. Cell 172(409–422):e421

Ivanova D, Zhelev Z, Getsov P, Nikolova B, Aoki I, Higashi T, Bakalova R (2018) Vitamin K: redox-modulation, prevention of mitochondrial dysfunction and anticancer effect. Redox Biol 16:352–358

Jacob RA, Pianalto FS, Agee RE (1992) Cellular ascorbate depletion in healthy men. J Nutr 122: 1111–1118

James ER, Carrell DT, Aston KI, Jenkins TG, Yeste M, Salas-Huetos A (2020) The role of the epididymis and the contribution of Epididymosomes to mammalian reproduction. Int J Mol Sci 21:5377

Janiszewski M, Do Carmo AO, Pedro MA, Silva E, Knobel E, Laurindo FR (2004) Platelet-derived exosomes of septic individuals possess proapoptotic NAD(P)H oxidase activity: A novel vascular redox pathway. Crit Care Med 32:818–825

Jankovičová J, Michalková K, Sečová P, Horovská Ľ, Antalíková J (2024) The extracellular vesicle tetraspanin CD63 journey from the testis through the epididymis to mature bull sperm. Sci Rep 14:29449

Jenkins AD, Lechene CP, Howards SS (1980) Concentrations of seven elements in the intraluminal fluids of the rat seminiferous tubules, rate testis, and epididymis. Biol Reprod 23:981–987

Jensen MB (2014) Vitamin D and male reproduction. Nat Rev Endocrinol 10:175–186

Jeppesen DK, Fenix AM, Franklin JL, Higginbotham JN, Zhang Q, Zimmerman LJ, Liebler DC, Ping J, Liu Q, Evans R et al (2019) Reassessment of exosome composition. Cell 177(428–445): e418

Jeppesen DK, Zhang Q, Franklin JL, Coffey RJ (2023) Extracellular vesicles and nanoparticles: emerging complexities. Trends Cell Biol 33:667–681

Jervis KM, Robaire B (2004) The effects of long-term vitamin E treatment on gene expression and oxidative stress damage in the aging Brown Norway rat epididymis. Biol Reprod 71:1088–1095

Jiang D, Li Z, Liu H, Liu H, Xia X, Xiang X (2024a) Plant exosome-like nanovesicles derived from sesame leaves as carriers for luteolin delivery: molecular docking, stability and bioactivity. Food Chem 438:137963

Jiang L, Yang F, Liao H, Chen W, Dai X, Peng C, Li Z, Wang H, Zhang T, Cao H (2024b) Molybdenum and cadmium cause blood–testis barrier dysfunction through ROS-mediated NLRP3 inflammasome activation in sheep. Sci Total Environ 906:167267

Jin S, Zhou F, Katirai F, Li PL (2011) Lipid raft redox signaling: molecular mechanisms in health and disease. Antioxid Redox Signal 15:1043–1083

Johansson S, Melhus H (2001) Vitamin A antagonizes calcium response to vitamin D in man. J Bone Miner Res 16:1899–1905

Jomova K, Valko M (2011) Advances in metal-induced oxidative stress and human disease. Toxicology 283:65–87

Jones DP (2008) Radical-free biology of oxidative stress. Am J Physiol Cell Physiol 295:C849–C868

Juarez-Rojas L, Casillas F, Lopez A, Betancourt M, Ommati MM, Retana-Marquez S (2021) Physiological role of reactive oxygen species in testis and epididymal spermatozoa. Andrologia 54:e14367

Juodeikis R, Jones E, Deery E, Beal DM, Stentz R, Krautler B, Carding SR, Warren MJ (2022) Nutrient smuggling: commensal gut bacteria-derived extracellular vesicles scavenge vitamin B12 and related cobamides for microbe and host acquisition. J Extracell Biol 1:e61

Jursa T, Smith DR (2009) Ceruloplasmin alters the tissue disposition and neurotoxicity of manganese, but not its loading onto transferrin. Toxicol Sci 107:182–193

Jyothidasan A, Sunny S, Murugesan S, Quiles JM, Challa AK, Dalley B, Cinghu SK, Nanda V, Rajasekaran NS (2022) Transgenic expression of Nrf2 induces a pro-reductive stress and adaptive cardiac remodeling in the mouse. Genes (Basel) 13:1514

Kaewma S, Namula Z, Nguyen ST, Lin Q, Torigoe N, Liu B, Nagahara M, Nii M, Taniguchi M, Otoi T (2024) Effects of ergothioneine supplementation on the quality of liquid-preserved and frozen-thawed boar semen. Acta Vet Hung 71:219–222

Kaido M, Mori K, Ide Y, Inoue N, Koide O (1991) Testicular damage by high doses of vitamin B6 (pyridoxine) in rats: a light and electron microscopical study. Exp Mol Pathol 55:63–82

Kalthur G, Raj S, Thiyagarajan A, Kumar S, Kumar P, Adiga SK (2011) Vitamin E supplementation in semen-freezing medium improves the motility and protects sperm from freeze-thaw-induced DNA damage. Fertil Steril 95:1149–1151

Kaneko T, Iuchi Y, Kobayashi T, Fujii T, Saito H, Kurachi H, Fujii J (2002) The expression of glutathione reductase in the male reproductive system of rats supports the enzymatic basis of glutathione function in spermatogenesis. Eur J Biochem 269:1570–1578

Kapustin AN, Davies JD, Reynolds JL, McNair R, Jones GT, Sidibe A, Schurgers LJ, Skepper JN, Proudfoot D, Mayr M et al (2011) Calcium regulates key components of vascular smooth muscle cell-derived matrix vesicles to enhance mineralization. Circ Res 109:e1–e12

Kapustin AN, Chatrou ML, Drozdov I, Zheng Y, Davidson SM, Soong D, Furmanik M, Sanchis P, De Rosales RT, Alvarez-Hernandez D et al (2015) Vascular smooth muscle cell calcification is mediated by regulated exosome secretion. Circ Res 116:1312–1323

Kavian N, Marut W, Servettaz A, Nicco C, Chereau C, Lemarechal H, Guilpain P, Chimini G, Galland F, Weill B et al (2015) Pantethine prevents murine systemic sclerosis through the inhibition of microparticle shedding. Arthritis Rheumatol 67:1881–1890

Keerthikumar S, Chisanga D, Ariyaratne D, Al Saffar H, Anand S, Zhao K, Samuel M, Pathan M, Jois M, Chilamkurti N et al (2016) ExoCarta: A web-based compendium of Exosomal cargo. J Mol Biol 428:688–692

Keller MA, Piedrafita G, Ralser M (2015) The widespread role of non-enzymatic reactions in cellular metabolism. Curr Opin Biotechnol 34:153–161

Kellokumpu S (2019) Golgi pH, ion and redox homeostasis: how much do they really matter? Front Cell Dev Biol 7:93

Keskes-Ammar L, Feki-Chakroun N, Rebai T, Sahnoun Z, Ghozzi H, Hammami S, Zghal K, Fki H, Damak J, Bahloul A (2003) Sperm oxidative stress and the effect of an oral vitamin E and selenium supplement on semen quality in infertile men. Arch Androl 49:83–94

Khan MM, Lee S, Couoh-Cardel S, Oot RA, Kim H, Wilkens S, Roh SH (2022) Oxidative stress protein Oxr1 promotes V-ATPase holoenzyme disassembly in catalytic activity-independent manner. EMBO J 41:e109360
Khatami PG, Soleimani A, Sharifi N, Aghadavod E, Asemi Z (2016) The effects of high-dose vitamin E supplementation on biomarkers of kidney injury, inflammation, and oxidative stress in patients with diabetic nephropathy: A randomized, double-blind, placebo-controlled trial. J Clin Lipidol 10:922–929
Kidroni G, Har-Nir R, Menezel J, Frutkoff IW, Palti Z, Ron M (1983) Vitamin D3 metabolites in rat epididymis: high 24,25-dihydroxy vitamin D3 levels in the cauda region. Biochem Biophys Res Commun 113:982–989
Kieliszek M, Bano I, Zare H (2021) A comprehensive review on selenium and its effects on human health and distribution in middle eastern countries. Biol Trace Elem Res 200:971
Kim B, Roy J, Shum WW, Da Silva N, Breton S (2015) Role of testicular luminal factors on basal cell elongation and proliferation in the mouse epididymis. Biol Reprod 92:9
Klotz LO, Sanchez-Ramos C, Prieto-Arroyo I, Urbanek P, Steinbrenner H, Monsalve M (2015) Redox regulation of FoxO transcription factors. Redox Biol 6:51–72
Krishnamoorthy L, Chang CJ (2018) Exosomal NADPH oxidase: delivering redox signaling for healing. Biochemistry 57:3993–3994
Kumar MA, Baba SK, Sadida HQ, Marzooqi SA, Jerobin J, Altemani FH, Algehainy N, Alanazi MA, Abou-Samra A-B, Kumar R et al (2024a) Extracellular vesicles as tools and targets in therapy for diseases. Signal Transduct Target Ther 9:27
Kumar N, Thorat ST, Kochewad SA, Reddy KS (2024b) Manganese nutrient mitigates ammonia, arsenic toxicity and high temperature stress using gene regulation via NFkB mechanism in fish. Sci Rep 14:1273
Lai Y, Reina-Gonzalez P, Maor G, Miller GW, Sarkar S (2023) Biotin rescues manganese-induced Parkinson's disease phenotypes and neurotoxicity. bioRxiv
Lakoma K, Kukharuk O, Sliz D (2023) The influence of metabolic factors and diet on fertility. Nutrients 15:1180
Lanska DJ (2012) The discovery of niacin, biotin, and pantothenic acid. Ann Nutr Metab 61:246
Lasser C (2019) Mapping Extracellular RNA sheds lights on distinct carriers. Cell 177:228–230
Lauw SJL, Ganguly R, Webster RD (2013) The electrochemical reduction of biotin (vitamin B7) and conversion into its ester. Electrochim Acta 114:514–520
Lebovitz RM, Zhang H, Vogel H, Cartwright J Jr, Dionne L, Lu N, Huang S, Matzuk MM (1996) Neurodegeneration, myocardial injury, and perinatal death in mitochondrial superoxide dismutase-deficient mice. Proc Natl Acad Sci USA 93:9782–9787
Lennicke C, Cochemé HM (2021) Redox metabolism: ROS as specific molecular regulators of cell signaling and function. Mol Cell 81:3691
Lennicke C, Rahn J, Lichtenfels R, Wessjohann LA, Seliger B (2015) Hydrogen peroxide – production, fate and role in redox signaling of tumor cells. Cell Commun Signal 13:39
Leone E, Mann T (1951) Ergothioneine in the seminal vesicle secretion. Nature 168:205–206
Leung GP, Tse CM, Chew SB, Wong PY (2001) Expression of multiple Na+/H+ exchanger isoforms in cultured epithelial cells from rat efferent duct and cauda epididymidis. Biol Reprod 64:482–490
Leung GP, Cheung KH, Leung CT, Tsang MW, Wong PY (2004) Regulation of epididymal principal cell functions by basal cells: role of transient receptor potential (Trp) proteins and cyclooxygenase-1 (COX-1). Mol Cell Endocrinol 216:5–13
Levine N, Marsh DJ (1971) Micropuncture studies of the electrochemical aspects of fluid and electrolyte transport in individual seminiferous tubules, the epididymis and the vas deferens in rats. J Physiol 213:557–570
Li Z, Wang S, Gong C, Hu Y, Liu J, Wang W, Chen Y, Liao Q, He B, Huang Y et al (2021) Effects of environmental and pathological hypoxia on male fertility. Front Cell Dev Biol 9:725933
Li W, Liang L, Liu S, Yi H, Zhou Y (2023) FSP1: a key regulator of ferroptosis. Trends Mol Med 29:753–764

Liang X, Zhang Q, Zhang Z, Ding Q, Mei X (2021) Improving stability of vitamin B5 through double salt formation. Cryst Growth Des 21:4997–5005

Lin YW, Nhieu J, Wei CW, Lin YL, Kagechika H, Wei LN (2021) Regulation of exosome secretion by cellular retinoic acid binding protein 1 contributes to systemic anti-inflammation. Cell Commun Signal 19:69

Lisovskaya TV, Dubrovina OS, Treshchilov IM, Senturina LB, Sevostyanova OY, Mayasina EN, Buev YE, Salimov DF (2021) Thyroid disorders and pathospermia in the ART clinic patients. Gynecol Endocrinol 37:4–7

Liu Y, Chipot C, Shao X, Cai W (2011) The effects of 7-dehydrocholesterol on the structural properties of membranes. Phys Biol 8:056005

Liu BY, Zhang BL, Gao DY, Li Q, Xu XY, Shum W (2022) Epididymal epithelial degeneration and lipid metabolism impairment account for male infertility in occludin knockout mice. Front Endocrinol (Lausanne) 13:1069319

Liu J, Hermo L, Ding D, Wei C, Mann JM, Yan X, Melnick AF, Wu Y, Withrow A, Cibelli J et al (2023) SYPL1 defines a vesicular pathway essential for sperm cytoplasmic droplet formation and male fertility. Nat Commun 14:5113

Lloyd D, Aon MA, Cortassa S (2001) Why homeodynamics, not homeostasis? ScientificWorldJournal 1:133–145

Lloyd-Evans E, Waller-Evans H (2020) Lysosomal ca(2+) homeostasis and signaling in health and disease, vol 12. Cold Spring Harb Perspect Biol, p 2867

Lobo V, Patil A, Phatak A, Chandra N (2010) Free radicals, antioxidants and functional foods: impact on human health. Pharmacogn Rev 4:118–126

Loftfield E, O'Connell CP, Abnet CC, Graubard BI, Liao LM, Beane Freeman LE, Hofmann JN, Freedman ND, Sinha R (2024) Multivitamin use and mortality risk in 3 prospective US cohorts. JAMA Netw Open 7:e2418729

Lotvall J, Hill AF, Hochberg F, Buzas EI, Di Vizio D, Gardiner C, Gho YS, Kurochkin IV, Mathivanan S, Quesenberry P et al (2014) Minimal experimental requirements for definition of extracellular vesicles and their functions: a position statement from the International Society for Extracellular Vesicles. J Extracell Vesicles 3:26913

Lowe NM (2021) The global challenge of hidden hunger: perspectives from the field. Proc Nutr Soc 80:283–289

Lu Q, Liang Q, Xi Y (2024) The effects of vitamin D supplementation on serum lipid profiles in people with type 2 diabetes: a systematic review and meta-analysis of randomized controlled trials. Front Nutr 11:1419747

Lupu VV, Lupu A, Jechel E, Starcea IM, Stoleriu G, Ioniuc I, Azoicai A, Danielescu C, Knieling A, Borka-Balas R et al (2024) The role of vitamin D in pediatric systemic lupus erythematosus – a double pawn in the immune and microbial balance. Front Immunol 15:1373904

Lyon P, Strippoli V, Fang B, Cimmino L (2020) B vitamins and one-carbon metabolism: implications in human health and disease. Nutrients 12

Ma H, Zhang BL, Liu BY, Shi S, Gao DY, Zhang TC, Shi HJ, Li Z, Shum WW (2019) Vitamin K2-dependent GGCX and MGP are required for homeostatic calcium regulation of sperm maturation. iScience 14:210–225

Maiorino M, Bosello V, Ursini F, Foresta C, Garolla A, Scapin M, Sztajer H, Flohe L (2003) Genetic variations of gpx-4 and male infertility in humans. Biol Reprod 68:1134–1141

Makarova J, Maltseva D, Tonevitsky A (2023) Challenges in characterization of transcriptomes of extracellular vesicles and non-vesicular extracellular RNA carriers. Front Mol Biosci 10: 1327985

Manford AG, Mena EL, Shih KY, Gee CL, McMinimy R, Martinez-Gonzalez B, Sherriff R, Lew B, Zoltek M, Rodriguez-Perez F et al (2021) Structural basis and regulation of the reductive stress response. Cell 184:5375–5390 e5316

Marin de Jesus S, Vigueras-Villasenor RM, Cortes-Barberena E, Hernandez-Rodriguez J, Montes S, Arrieta-Cruz I, Perez-Aguirre SG, Bonilla-Jaime H, Limon-Morales O, Arteaga-

Silva M (2024) Zinc and its impact on the function of the testicle and epididymis. Int J Mol Sci 25:8991

Martin-DeLeon PA (2006) Epididymal SPAM1 and its impact on sperm function. Mol Cell Endocrinol 250:114–121

Martin-DeLeon PA (2015) Epididymosomes: transfer of fertility-modulating proteins to the sperm surface. Asian J Androl 17:720–725

Martinez A, Rodriguez-Girones MA, Barbosa A, Costas M (2008) Donator acceptor map for carotenoids, melatonin and vitamins. J Phys Chem A 112:9037–9042

Matorras R, Perez-Sanz J, Corcostegui B, Perez-Ruiz I, Malaina I, Quevedo S, Aspichueta F, Crisol L, Martinez-Indart L, Prieto B et al (2020) Effect of vitamin E administered to men in infertile couples on sperm and assisted reproduction outcomes: a double-blind randomized study. F S Rep 1:219–226

Mazzilli R, Medenica S, Di Tommaso AM, Fabozzi G, Zamponi V, Cimadomo D, Rienzi L, Ubaldi FM, Watanabe M, Faggiano A et al (2023) The role of thyroid function in female and male infertility: a narrative review. J Endocrinol Investig 46:15–26

Mear LO, Tsai PS, Tamessar CT, Schjenken JE, Nixon B (2024) Epididymosomes: composition and functions for sperm maturation. Adv Anat Embryol Cell Biol. https://doi.org/10.1007/102_2024_7

Mendel RR, Kruse T (2012) Cell biology of molybdenum in plants and humans. Biochim Biophys Acta 1823:1568–1579

Menditto A, Pietraforte D, Minetti M (1997) Ascorbic acid in human seminal plasma is protected from iron-mediated oxidation, but is potentially exposed to copper-induced damage. Hum Reprod 12:1699–1705

Michel V, Duan Y, Stoschek E, Bhushan S, Middendorff R, Young JM, Loveland KL, Kretser DM, Hedger MP, Meinhardt A (2016) Uropathogenic Escherichia coli causes fibrotic remodelling of the epididymis. J Pathol 240:15–24

Michniewicz F, Saletta F, Rouaen JRC, Hewavisenti RV, Mercatelli D, Cirillo G, Giorgi FM, Trahair T, Ziegler D, Vittorio O (2021) Copper: an intracellular Achilles' heel allowing the targeting of epigenetics, kinase pathways, and cell metabolism in cancer therapeutics. ChemMedChem 16:2315–2329

Miner GE, Sullivan KD, Zhang C, Hurst LR, Starr ML, Rivera-Kohr DA, Jones BC, Guo A, Fratti RA (2019) Copper blocks V-ATPase activity and SNARE complex formation to inhibit yeast vacuole fusion. Traffic 20:841

Minich WB (2022) Selenium metabolism and biosynthesis of Selenoproteins in the human body. Biochemistry (Mosc) 87:S168–S102

Mirnamniha M, Faroughi F, Tahmasbpour E, Ebrahimi P, Beigi Harchegani A (2019) An overview on role of some trace elements in human reproductive health, sperm function and fertilization process. Rev Environ Health 34:339–348

Mishima E, Ito J, Wu Z, Nakamura T, Wahida A, Doll S, Tonnus W, Nepachalovich P, Eggenhofer E, Aldrovandi M et al (2022) A non-canonical vitamin K cycle is a potent ferroptosis suppressor. Nature 608:778–783

Mishima E, Wahida A, Seibt T, Conrad M (2023) Diverse biological functions of vitamin K: from coagulation to ferroptosis. Nat Metab 5:924–932

Misko A, Mahtani K, Abbott J, Schwarz G, Atwal P (1993) Molybdenum cofactor deficiency. In: Adam MP, Feldman J, Mirzaa GM, Pagon RA, Wallace SE, Amemiya A (eds) GeneReviews ((R)). Seattle (WA)

Mizumoto C, Kawabata H, Uchiyama T, Sakamoto S, Kanda J, Tomosugi N, Takaori-Kondo A (2012) Acidic milieu augments the expression of hepcidin, the central regulator of iron homeostasis. Int J Hematol 96:701–709

Mohammadi H, Golbabaei F, Dehghan SF, Imani H, Ramezani Tehrani F, Khodakarim Ardakani S (2022) The influence of vitamin E and Omega-3 fatty acids on reproductive health indices among male workers exposed to electromagnetic fields. Am J Mens Health 16:15579883221074821

Momeni HR, Eskandari N (2012) Effect of vitamin E on sperm parameters and DNA integrity in sodium arsenite-treated rats. Iran J Reprod Med 10:249–256

Moon N, Morgan CP, Marx-Rattner R, Jeng A, Johnson RL, Chikezie I, Mannella C, Sammel MD, Epperson CN, Bale TL (2024) Stress increases sperm respiration and motility in mice and men. Nat Commun 15:7900

Moustakli E, Zikopoulos A, Skentou C, Katopodis P, Domali E, Potiris A, Stavros S, Zachariou A (2024) Impact of reductive stress on human infertility: underlying mechanisms and perspectives. Int J Mol Sci 25:11802

Mrowicka M, Mrowicki J, Dragan G, Majsterek I (2023) The importance of thiamine (vitamin B1) in humans. Biosci Rep 43:BSR20230374

Mu N, Li J, Zeng L, You J, Li R, Qin A, Liu X, Yan F, Zhou Z (2023) Plant-derived exosome-like Nanovesicles: current Progress and prospects. Int J Nanomedicine 18:4987–5009

Mumtaz N, Akhtar MF, Saleem A, Riaz A (2022) Harmful consequences of proton pump inhibitors on male fertility: an evidence from subchronic toxicity study of esomeprazole and lansoprazole in Wistar rats. Int J Endocrinol 2022:4479261

Murillo OD, Thistlethwaite W, Rozowsky J, Subramanian SL, Lucero R, Shah N, Jackson AR, Srinivasan S, Chung A, Laurent CD et al (2019) exRNA atlas analysis reveals distinct Extracellular RNA cargo types and their carriers present across human biofluids. Cell 177:463–477 e415

Murshed M, Schinke T, McKee MD, Karsenty G (2004) Extracellular matrix mineralization is regulated locally; different roles of two gla-containing proteins. J Cell Biol 165:625–630

Musset B, Clark RA, DeCoursey TE, Petheo GL, Geiszt M, Chen Y, Cornell JE, Eddy CA, Brzyski RG, El Jamali A (2012) NOX5 in human spermatozoa: expression, function, and regulation. J Biol Chem 287:9376–9388

Mutalip, S.S.M. (2019). Vitamin E and reproductive health. In Molecular Nutrition

Myers C, Cornwall GA (2024) Host defense amyloids: biosensors of the immune system? Andrology 12:973–980

Najafi A, Kia HD, Mohammadi H, Najafi MH, Zanganeh Z, Sharafi M, Martinez-Pastor F, Adeldust H (2014) Different concentrations of cysteamine and ergothioneine improve microscopic and oxidative parameters in ram semen frozen with a soybean lecithin extender. Cryobiology 69:68–73

Nakase I, Ueno N, Matsuzawa M, Noguchi K, Hirano M, Omura M, Takenaka T, Sugiyama A, Bailey Kobayashi N, Hashimoto T et al (2021) Environmental pH stress influences cellular secretion and uptake of extracellular vesicles. FEBS Open Bio 11:753–767

Negri M, Amatrudo F, Gentile A, Patalano R, Monto T, de Angelis C, Simeoli C, Pirchio R, Auriemma RS, Colao A et al (2022) Vitamin D reverts the exosome-mediated transfer of cancer resistance to the mTOR inhibitor Everolimus in hepatocellular carcinoma. Front Oncol 12: 874091

Niess JH, Reinecker HC (2006) Dendritic cells: the commanders-in-chief of mucosal immune defenses. Curr Opin Gastroenterol 22:354–360

Nixon B, De Iuliis GN, Hart HM, Zhou W, Mathe A, Bernstein IR, Anderson AL, Stanger SJ, Skerrett-Byrne DA, Jamaluddin MFB et al (2018) Proteomic profiling of mouse Epididymosomes reveals their contributions to post-testicular sperm maturation. Mol Cell Proteomics 18:S91–s108

Nonogaki T, Noda Y, Narimoto K, Shiotani M, Mori T, Matsuda T, Yoshida O (1992) Localization of Cuzn-superoxide dismutase in the human male genital organs. Hum Reprod 7:81–85

Nunes P, Demaurex N (2014) Redox regulation of store-operated Ca2+ entry. Antioxid Redox Signal 21:915–932

Offringa AK, Bourgonje AR, Schrier MS, Deth RC, van Goor H (2021) Clinical implications of vitamin B(12) as redox-active cofactor. Trends Mol Med 27:931–934

O'Flaherty C (2019) Orchestrating the antioxidant defenses in the epididymis. Andrology 7:662–668

Ogorek M, Herman S, Pierzchala O, Bednarz A, Rajfur Z, Baster Z, Grzmil P, Starzynski RR, Szudzik M, Jonczy A et al (2019) Molecular machinery providing copper bioavailability for spermatozoa along the epididymial tubule in mouse. Biol Reprod 100:1505–1520

Oleszczuk K, Augustinsson L, Bayat N, Giwercman A, Bungum M (2013) Prevalence of high DNA fragmentation index in male partners of unexplained infertile couples. Andrology 1:357–360
Oliva R (2006) Protamines and male infertility. Hum Reprod Update 12:417–435
Olli KE, Li K, Galileo DS, Martin-DeLeon PA (2018) Plasma membrane calcium ATPase 4 (PMCA4) co-ordinates calcium and nitric oxide signaling in regulating murine sperm functional activity. J Cell Physiol 233:11–22
Ong DE, Newcomer ME, Lareyre JJ, Orgebin-Crist MC (2000) Epididymal retinoic acid-binding protein. Biochim Biophys Acta 1482:209–217
Orgebin-Crist MC (1967) Sperm maturation in rabbit epididymis. Nature 216:816–818
Orgebin-Crist MC (1969) Studies on the function of the epididymis. Biol Reprod 1(Suppl 1): 155–175
Orgebin-Crist MC, Lareyre JJ, Suzuki K, Araki Y, Fouchecourt S, Matusik RJ, Ong DE (2002) Retinoids and epididymal function. In: Robaire B, Hinton BT (eds) The epididymis: from molecular to clinical practice A comprehensive survey of the efferent ducts, the epididymis and the vas deferens. Kluwer Academic/Plenum, New York, pp 339–352
Ornoy A, Goodwin D, Noff D, Edelstein S (1978) 24, 25-Dihydroxyvitamin D is a metabolite of vitamin D essential for bone formation. Nature 276:517–519
Oshikawa S, Sonoda H, Ikeda M (2016) Aquaporins in urinary Extracellular vesicles (exosomes). Int J Mol Sci 17
Oshikawa-Hori S, Yokota-Ikeda N, Sonoda H, Ikeda M (2019) Urinary extracellular vesicular release of aquaporins in patients with renal transplantation. BMC Nephrol 20:216
Owen KN, Dewald O (2025) Vitamin E toxicity. In StatPearls (Treasure Island (FL) ineligible companies. Disclosure: Olga Dewald declares no relevant financial relationships with ineligible companies.)
Palladino MA, Powell JD, Korah N, Hermo L (2004) Expression and localization of hypoxia-inducible factor-1 subunits in the adult rat epididymis. Biol Reprod 70:1121–1130
Panner Selvam MK, Agarwal A, Henkel R, Finelli R, Robert KA, Iovine C, Baskaran S (2020) The effect of oxidative and reductive stress on semen parameters and functions of physiologically normal human spermatozoa. Free Radic Biol Med 152:375–385
Panth N, Gavarkovs A, Tamez M, Mattei J (2018) The influence of diet on fertility and the implications for public health nutrition in the United States. Front Public Health 6:211
Park YJ, Battistone MA, Kim B, Breton S (2017) Relative contribution of clear cells and principal cells to luminal pH in the mouse epididymis. Biol Reprod 96:366–375
Parra M, Stahl S, Hellmann H (2018) Vitamin B(6) and its role in cell metabolism and physiology. Cells 7(7). https://doi.org/10.3390/cells7070084
Partal-Lorente AB, Maldonado-Ezequiel V, Martinez-Navarro L, Herrera-Contreras I, Gutierrez-Repiso C, García-Fuentes E, Santiago-Fernandez P (2017) Iodine is associated to semen quality in men who undergo consultations for infertility. Reprod Toxicol 73:1–7
Pastor-Soler N, Pietrement C, Breton S (2005) Role of acid/base transporters in the male reproductive tract and potential consequences of their malfunction. Physiology (Bethesda) 20:417–428
Patel R, Al-Dossary AA, Stabley DL, Barone C, Galileo DS, Strehler EE, Martin-DeLeon PA (2013) Plasma membrane Ca2+-ATPase 4 in murine epididymis: secretion of splice variants in the luminal fluid and a role in sperm maturation. Biol Reprod 89:6
Paul BD (2022) Ergothioneine: A stress vitamin with antiaging, vascular, and neuroprotective roles? Antioxid Redox Signal 36:1306–1317
Paul N, Talluri TR, Nag P, Kumaresan A (2021) Epididymosomes: A potential male fertility influencer. Andrologia 53:e14155
Paunescu TG, Shum WW, Huynh C, Lechner L, Goetze B, Brown D, Breton S (2014) High-resolution helium ion microscopy of epididymal epithelial cells and their interaction with spermatozoa. Mol Hum Reprod 20:929–937

Payet T, Astier J, Bournot L, Sicard F, Robert S, Lacroix R, Wabitsch M, Landrier JF, Mounien L (2025) Vitamin D modulates the content of inflammatory microRNAs in extracellular vesicles from human adipocyte cells in inflammatory context. Biofactors 51:e70003

Pecora G, Sciarra F, Gangitano E, Venneri MA-O (2023) How food choices impact on male fertility, vol 12. Curr Nutr Rep, p 864

Penet MF, Abou-Hamdan M, Coltel N, Cornille E, Grau GE, de Reggi M, Gharib B (2008) Protection against cerebral malaria by the low-molecular-weight thiol pantethine. Proc Natl Acad Sci USA 105:1321–1326

Perez-Torres I, Guarner-Lans V, Rubio-Ruiz ME (2017) Reductive stress in inflammation-associated diseases and the pro-oxidant effect of antioxidant agents. Int J Mol Sci 18:2098

Perut F, Roncuzzi L, Avnet S, Massa A, Zini N, Sabbadini S, Giampieri F, Mezzetti B, Baldini N (2021) Strawberry-derived exosome-like nanoparticles prevent oxidative stress in human mesenchymal stromal cells. Biomolecules 11:87

Pfeifer H, Conrad M, Roethlein D, Kyriakopoulos A, Brielmeier M, Bornkamm GW, Behne D (2001) Identification of a specific sperm nuclei selenoenzyme necessary for protamine thiol cross-linking during sperm maturation. FASEB J 15:1236–1238

Pleuger C, Silva EJR, Pilatz A, Bhushan S, Meinhardt A (2020) Differential immune response to infection and acute inflammation along the epididymis. Front Immunol 11:599594

Pront R, Margalioth EJ, Green R, Eldar-Geva T, Maimoni Z, Zimran A, Elstein D (2009) Prevalence of low serum cobalamin in infertile couples. Andrologia 41:46–50

Pruneda A, Yeung CH, Bonet S, Pinart E, Cooper TG (2007) Concentrations of carnitine, glutamate and myo-inositol in epididymal fluid and spermatozoa from boars. Anim Reprod Sci 97:344–355

Punab M, Poolamets O, Paju P, Vihljajev V, Pomm K, Ladva R, Korrovits P, Laan M (2017) Causes of male infertility: a 9-year prospective monocentre study on 1737 patients with reduced total sperm counts. Hum Reprod 32:18–31

Qazi IH, Angel C, Yang H, Zoidis E, Pan B, Wu Z, Ming Z, Zeng CJ, Meng Q, Han H et al (2019) Role of selenium and Selenoproteins in male reproductive function: A review of past and present evidences. Antioxidants (Basel) 8

Quesenberry PJ, Aliotta J, Camussi G, Abdel-Mageed AB, Wen S, Goldberg L, Zhang HG, Tetta C, Franklin J, Coffey RJ et al (2015) Potential functional applications of extracellular vesicles: a report by the NIH common fund Extracellular RNA communication consortium. J Extracell Vesicles 4:27575

Radkhah N, Zarezadeh M, Jamilian P, Ostadrahimi A (2023) The effect of vitamin D supplementation on lipid profiles: an umbrella review of meta-analyses. Adv Nutr 14:1479–1498

Randaccio L, Geremia S, Demitri N, Demitri N, Wuerges J, Wuerges J (2010) Vitamin B12: unique metalorganic compounds and the most complex vitamins. Molecules 15:3228

Ranganathan S, Jackson RL, Harmony JA (1982) Effect of pantethine on the biosynthesis of cholesterol in human skin fibroblasts. Atherosclerosis 44:261–273

Ranjan R, Kumar M, Swain DK, Singh SP, Kharche SD, Chauhan MS (2022) Vitamin B7 protects DNA damage and stabilizes mitochondrial transmembrane potential from cryoinjury. Small Rumin Res 212:106719

Ravi, Kumar A, Bhattacharyya S, Singh J (2023) Thiol reductive stress activates the hypoxia response pathway. EMBO J 42:e114093

Raviprolu VT, Gregory A, Banda I, McArthur SG, McArthur SE, Goddard WA 3rd, Musgrave CB 3rd, Lavallo V (2025) Confirmation of Breslow's hypothesis: a carbene stable in liquid water. Sci Adv 11:eadr9681

Reay WR, Kiltschewskij DJ, Di Biase MA, Gerring ZF, Kundu K, Surendran P, Greco LA, Clarke ED, Collins CE, Mondul AM et al (2024) Genetic influences on circulating retinol and its relationship to human health. Nat Commun 15:1490

Reiss J, Hahnewald R (2011) Molybdenum cofactor deficiency: mutations in GPHN, MOCS1, and MOCS2. Hum Mutat 32:10–18

Rejraji H, Sion B, Prensier G, Carreras M, Motta C, Frenoux JM, Vericel E, Grizard G, Vernet P, Drevet JR (2006) Lipid remodeling of murine epididymosomes and spermatozoa during epididymal maturation. Biol Reprod 74:1104–1113
Rengan AK, Agarwal A, van der Linde M, du Plessis SS (2012) An investigation of excess residual cytoplasm in human spermatozoa and its distinction from the cytoplasmic droplet. Reprod Biol Endocrinol 10:92
Reynolds JL, Joannides AJ, Skepper JN, McNair R, Schurgers LJ, Proudfoot D, Jahnen-Dechent W, Weissberg PL, Shanahan CM (2004) Human vascular smooth muscle cells undergo vesicle-mediated calcification in response to changes in extracellular calcium and phosphate concentrations: a potential mechanism for accelerated vascular calcification in ESRD. J Am Soc Nephrol 15:2857–2867
Rinaldi VD, Donnard E, Gellatly K, Rasmussen M, Kucukural A, Yukselen O, Garber M, Sharma U, Rando OJ (2020) An atlas of cell types in the mouse epididymis and vas deferens. elife 9
Rishavy MA, Berkner KL (2012) Vitamin K oxygenation, glutamate carboxylation, and processivity: defining the three critical facets of catalysis by the vitamin K-dependent carboxylase. Adv Nutr 3:135–148
Robaire B, Hinton BT (2015) The epididymis. In: Plant TM, Zeleznik AJ (eds) Knobil and Neill's physiology of reproduction. Academic, Elsevier Inc., pp 691–771
Rodrigues A, Queiroz DB, Honda L, Silva EJ, Hall SH, Avellar MC (2008) Activation of toll-like receptor 4 (TLR4) by in vivo and in vitro exposure of rat epididymis to lipopolysaccharide from Escherichia Coli. Biol Reprod 79:1135–1147
Rolf C, Cooper TG, Yeung CH, Nieschlag E (1999) Antioxidant treatment of patients with asthenozoospermia or moderate oligoasthenozoospermia with high-dose vitamin C and vitamin E: a randomized, placebo-controlled, double-blind study. Hum Reprod 14:1028–1033
Roman-Garcia P, Quiros-Gonzalez I, Mottram L, Lieben L, Sharan K, Wangwiwatsin A, Tubio J, Lewis K, Wilkinson D, Santhanam B et al (2014) Vitamin B12-dependent taurine synthesis regulates growth and bone mass. J Clin Invest 124:2988–3002
Rousset B, Dupuy C, Miot F (2015) Chapter 2 thyroid hormone synthesis and secretion. MDText.com, Inc., South Dartmouth (MA)
Sadeghi N, Boissonneault G, Tavalaee M, Nasr-Esfahani MH (2023) Oxidative versus reductive stress: a delicate balance for sperm integrity. Syst Biol Reprod Med 69:20–31
Santolini J, Wootton SA, Jackson AA, Feelisch M (2019) The redox architecture of physiological function. Curr Opin Physio 9:34–47
Schafer FQ, Buettner GR (2001) Redox environment of the cell as viewed through the redox state of the glutathione disulfide/glutathione couple. Free Radical Bio Med 30:1191–1212
Schnorr CE, Morrone Mda S, Weber MH, Lorenzi R, Behr GA, Moreira JC (2011) The effects of vitamin A supplementation to rats during gestation and lactation upon redox parameters: increased oxidative stress and redox modulation in mothers and their offspring. Food Chem Toxicol 49:2645–2654
Schumacher W, Holliger C, Zehnder AJ, Hagen WR (1997) Redox chemistry of cobalamin and iron-sulfur cofactors in the tetrachloroethene reductase of Dehalobacter restrictus. FEBS Lett 409:421–425
Schurgers LJ, Cranenburg EC, Vermeer C (2008) Matrix Gla-protein: the calcification inhibitor in need of vitamin K. Thromb Haemost 100:593–603
Schurgers LJ, Uitto J, Reutelingsperger CP (2013) Vitamin K-dependent carboxylation of matrix Gla-protein: a crucial switch to control ectopic mineralization. Trends Mol Med 19:217–226
Schwarz G (2016) Molybdenum cofactor and human disease. Curr Opin Chem Biol 31:179–187
Seidel T, Scholl S, Krebs M, Rienmuller F, Marten I, Hedrich R, Hanitzsch M, Janetzki P, Dietz KJ, Schumacher K (2012) Regulation of the V-type ATPase by redox modulation. Biochem J 448: 243–251

Seligman J, Newton GL, Fahey RC, Shalgi R, Kosower NS (2005) Nonprotein thiols and disulfides in rat epididymal spermatozoa and epididymal fluid: role of gamma-glutamyl-transpeptidase in sperm maturation. J Androl 26:629–637; discussion 638-640

Sengupta P, Dutta S, Alahmar AT (2022) Reductive stress and male infertility. Adv Exp Med Biol 1391:311–321

Sentman ML, Granstrom M, Jakobson H, Reaume A, Basu S, Marklund SL (2006) Phenotypes of mice lacking extracellular superoxide dismutase and copper- and zinc-containing superoxide dismutase. J Biol Chem 281:6904–6909

Shanmugam G, Wang D, Gounder SS, Fernandes J, Litovsky SH, Whitehead K, Radhakrishnan RK, Franklin S, Hoidal JR, Kensler TW et al (2020) Reductive stress causes pathological cardiac remodeling and diastolic dysfunction. Antioxid Redox Signal 32:1293–1312

Sharma U, Conine CC, Shea JM, Boskovic A, Derr AG, Bing XY, Belleannee C, Kucukural A, Serra RW, Sun F et al (2016) Biogenesis and function of tRNA fragments during sperm maturation and fertilization in mammals. Science 351:391–396

Sharma U, Sun F, Conine CC, Reichholf B, Kukreja S, Herzog VA, Ameres SL, Rando OJ (2018) Small RNAs are trafficked from the epididymis to developing mammalian sperm. Dev Cell 46: 481–494 e486

Shen J, Griffiths PT, Campbell SJ, Utinger B, Kalberer M, Paulson SE (2021) Ascorbate oxidation by iron, copper and reactive oxygen species: review, model development, and derivation of key rate constants. Sci Rep 11:7417

Sheweita SA, El-Dafrawi YA, El-Ghalid OA, Ghoneim AA, Wahid A (2022) Antioxidants (selenium and garlic) alleviated the adverse effects of tramadol on the reproductive system and oxidative stress markers in male rabbits. Sci Rep 12:13958

Shi J, Fok KL, Dai P, Qiao F, Zhang M, Liu H, Sang M, Ye M, Liu Y, Zhou Y et al (2021) Spatio-temporal landscape of mouse epididymal cells and specific mitochondria-rich segments defined by large-scale single-cell RNA-seq. Cell Discov 7:34

Shum WW, Da Silva N, McKee M, Smith PJ, Brown D, Breton S (2008) Transepithelial projections from basal cells are luminal sensors in pseudostratified epithelia. Cell 135:1108–1117

Shum WW, Da Silva N, Brown D, Breton S (2009) Regulation of luminal acidification in the male reproductive tract via cell-cell crosstalk. J Exp Biol 212:1753–1761

Shum WW, Da Silva N, Belleannee C, McKee M, Brown D, Breton S (2011a) Regulation of V-ATPase recycling via a RhoA- and ROCKII-dependent pathway in epididymal clear cells. Am J Physiol Cell Physiol 301:C31–C43

Shum WW, Ruan YC, Da Silva N, Breton S (2011b) Establishment of cell-cell crosstalk in the epididymis: control of luminal acidification. J Androl 32:576

Shum W, Zhang BL, Cao AS, Zhou X, Shi SM, Zhang ZY, Gu LY, Shi S (2022) Calcium homeostasis in the Epididymal microenvironment: is Extracellular calcium a cofactor for matrix Gla protein-dependent scavenging regulated by vitamins. Front Cell Dev Biol 10:827940

Siatka T, Mat'us M, Moravcova M, Harcarova P, Lomozova Z, Matousova K, Suwanvecho C, Krcmova LK, Mladenka P (2025) Biological, dietetic and pharmacological properties of vitamin B(9). NPJ Sci Food 9:30

Sies H, Belousov VV, Chandel NS, Davies MJ, Jones DP, Mann GE, Murphy MP, Yamamoto M, Winterbourn C (2022) Defining roles of specific reactive oxygen species (ROS) in cell biology and physiology. Nat Rev Mol Cell Biol 23:499–515

Sies H, Mailloux RJ, Jakob U (2024) Fundamentals of redox regulation in biology. Nat Rev Mol Cell Biol 25:701–719

Siragy HM, Carey RM (1997) The subtype 2 (AT2) angiotensin receptor mediates renal production of nitric oxide in conscious rats. J Clin Invest 100:264–269

Skakkebaek NE, Lindahl-Jacobsen R, Levine H, Andersson AM, Jorgensen N, Main KM, Lidegaard O, Priskorn L, Holmboe SA, Brauner EV et al (2022) Environmental factors in declining human fertility. Nat Rev Endocrinol 18:139–157

Skerget S, Rosenow MA, Petritis K, Karr TL (2015) Sperm proteome maturation in the mouse epididymis. PLoS One 10:e0140650

Skerrett-Byrne DA, Anderson AL, Bromfield EG, Bernstein IR, Mulhall JE, Schjenken JE, Dun MD, Humphrey SJ, Nixon B (2022) Global profiling of the proteomic changes associated with the post-testicular maturation of mouse spermatozoa. Cell Rep 41:111655
Skog J, Würdinger T, van Rijn S, van Rijn S, Meijer DH, Gainche L, Sena-Esteves M, Curry WT Jr, Carter BS, Krichevsky AM, Breakefield XO (2008) Glioblastoma microvesicles transport RNA and proteins that promote tumour growth and provide diagnostic biomarkers. Nat Cell Biol 10: 1470
Skoracka K, Eder PA-O, Łykowska-Szuber L, Dobrowolska AA-O, Krela-Kaźmierczak IA-O (2020) Diet and nutritional factors in male (in)fertility-underestimated factors. J Clin Med 9. https://doi.org/10.3390/jcm9051400
Smith DC, Klebanoff SJ (1970) A uterine fluid-mediated sperm-inhibitory system. Biol Reprod 3: 229–235
Solier S, Müller S, Cañeque T, Versini A, Mansart A, Sindikubwabo F, Baron L, Emam L, Gestraud P, Pantoş GD et al (2023) A druggable copper-signalling pathway that drives inflammation. Nature 617:386–394
Song Y, Mao C, Zhong Q, Zhang R, Jiang D, Sun X (2024) Role of hydrogen sulfide in the male reproductive system. Front Endocrinol (Lausanne) 15:1377090
Sonmez M, Turk G, Yuce A (2005) The effect of ascorbic acid supplementation on sperm quality, lipid peroxidation and testosterone levels of male Wistar rats. Theriogenology 63:2063–2072
Sprenger HG, Mittenbuhler MJ, Sun Y, Van Vranken JG, Schindler S, Jayaraj A, Khetarpal SA, Smythers AL, Vargas-Castillo A, Puszynska AM et al (2025) Ergothioneine controls mitochondrial function and exercise performance via direct activation of MPST. Cell Metab 37:857–869 e859
Stafford DW (2005) The vitamin K cycle. J Thromb Haemost 3:1873–1878
Stammler A, Hau T, Bhushan S, Meinhardt A, Jonigk D, Lippmann T, Pilatz A, Schneider-Huther I, Middendorff R (2015) Epididymitis: ascending infection restricted by segmental boundaries. Hum Reprod 30:1557–1565
Stockwell BR, Friedmann Angeli JP, Bayir H, Bush AI, Conrad M, Dixon SJ, Fulda S, Gascon S, Hatzios SK, Kagan VE et al (2017) Ferroptosis: A regulated cell death nexus linking metabolism, redox biology, and disease. Cell 171:273–285
Stoltenberg M, Ernst E, Andreasen A, Danscher G (1996) Histochemical localization of zinc ions in the epididymis of the rat. Histochem J 28:173
Studer JM, Schweer WP, Gabler NK, Ross JW (2022) Functions of manganese in reproduction. Anim Reprod Sci 238:106924
Suarez-Moreira E, Yun J, Birch CS, Williams JH, McCaddon A, Brasch NE (2009) Vitamin B (12) and redox homeostasis: cob(II)alamin reacts with superoxide at rates approaching superoxide dismutase (SOD). J Am Chem Soc 131:15078–15079
Sullivan R (2008) Principal cells' apocrine secretion of Epididymosomes and its function in sperm maturation. Biol Reprod 78:90–90
Sullivan R (2015) Epididymosomes: a heterogeneous population of microvesicles with multiple functions in sperm maturation and storage. Asian J Androl 17:726–729
Sullivan R, Mieusset R (2016) The human epididymis: its function in sperm maturation. Hum Reprod Update 22:574–587
Sullivan R, Saez F (2013) Epididymosomes, prostasomes, and liposomes: their roles in mammalian male reproductive physiology. Reproduction 146:R21–R35
Sullivan R, Saez F, Girouard J, Frenette G (2005) Role of exosomes in sperm maturation during the transit along the male reproductive tract. Blood Cells Mol Dis 35:1–10
Sullivan R, Frenette G, Girouard J (2007) Epididymosomes are involved in the acquisition of new sperm proteins during epididymal transit. Asian J Androl 9:483–491
Sullivan R, Legare C, Lamontagne-Proulx J, Breton S, Soulet D (2019) Revisiting structure/functions of the human epididymis. Andrology 7:748–757
Sun H-J, Rathinasabapathi B, Wu B, Luo J, Pu L-P, Ma LQ (2014a) Arsenic and selenium toxicity and their interactive effects in humans. Environ Int 69:148–158

Sun Y, Zhang WJ, Zhao X, Yuan RP, Jiang H, Pu XP (2014b) PARK7 protein translocating into spermatozoa mitochondria in Chinese asthenozoospermia. Reproduction 148:249–257

Sun Y, Chen C, Liu GG, Wang M, Shi C, Yu G, Lv F, Wang N, Zhang S (2020) The association between iodine intake and semen quality among fertile men in China. BMC Public Health 20:461

Tabatabaei Vakili S, Zeidi R, Nabhani R (2025) Protective effect of B1 and B12 vitamins on post-thaw sperm quality parameters and seminal plasma antioxidant status in rams. Vet Med Sci 11:e70241

Tan YL, Sou NL, Tang FY, Ko HA, Yeh WT, Peng JH, Chiang EI (2020) Tracing metabolic fate of mitochondrial glycine cleavage system derived Formate in vitro and in vivo. Int J Mol Sci 21:8808

Tarrago MG, Chini CCS, Kanamori KS, Warner GM, Caride A, de Oliveira GC, Rud M, Samani A, Hein KZ, Huang R et al (2018) A potent and specific CD38 inhibitor ameliorates age-related metabolic dysfunction by reversing tissue NAD(+) decline. Cell Metab 27:1081–1095 e1010

Taylor A, Robson A, Houghton BC, Jepson CA, Ford WC, Frayne J (2013) Epididymal specific, selenium-independent GPX5 protects cells from oxidative stress-induced lipid peroxidation and DNA mutation. Hum Reprod 28:2332–2342

Thimon V, Frenette G, Saez F, Thabet M, Sullivan R (2008) Protein composition of human epididymosomes collected during surgical vasectomy reversal: a proteomic and genomic approach. Hum Reprod 23:1698–1707

Tkach M, Thery C (2016) Communication by Extracellular vesicles: where we are and where we need to go. Cell 164:1226–1232

Tomar A, Gomez-Velazquez M, Gerlini R, Comas-Armangue G, Makharadze L, Kolbe T, Boersma A, Dahlhoff M, Burgstaller JP, Lassi M et al (2024) Epigenetic inheritance of diet-induced and sperm-borne mitochondrial RNAs. Nature 630:720–727

Tramer F, Rocco F, Micali F, Sandri G, Panfili E (1998) Antioxidant systems in rat epididymal spermatozoa. Biol Reprod 59:753–758

Trigg NA, Stanger SJ, Zhou W, Skerrett-Byrne DA, Sipila P, Dun MD, Eamens AL, De Iuliis GN, Bromfield EG, Roman SD et al (2021) A novel role for milk fat globule-EGF factor 8 protein (MFGE8) in the mediation of mouse sperm-extracellular vesicle interactions. Proteomics 21:e2000079

Truong DH, Eghbal MA, Hindmarsh W, Roth SH, O'Brien PJ (2006) Molecular mechanisms of hydrogen sulfide toxicity. Drug Metab Rev 38:733–744

Tsang T, Davis CI, Brady DC (2021) Copper biology. Curr Biol 31:R421–R427

Tsvetkov P, Coy S, Petrova B, Dreishpoon M, Verma A, Abdusamad M, Rossen J, Joesch-Cohen L, Humeidi R, Spangler RD et al (2022) Copper induces cell death by targeting lipoylated TCA cycle proteins. Science 375:1254–1261

Tully CA, Alesi S, McPherson NO, Sharkey DJ, Teong XT, Tay CT, Silva TR, Puglisi C, Barsby JP, Moran LJ et al (2024) Assessing the influence of preconception diet on male fertility: a systematic scoping review. Hum Reprod Update 30:243

Turner TT (2002) Necessity's potion: inorganic ions and small organic molecules in the epididymial lumen. In: Robaire BH, Hinton BT (eds) The epididymis: from molecules to clinical practice. Kluwer Academic/Plenum Publishers, New York, pp 131–150

Tvrda E, Peer R, Sikka SC, Agarwal A (2015) Iron and copper in male reproduction: a double-edged sword. J Assist Reprod Genet 32:3–16

Unsal V, Cicek M, Sabancilar I (2021) Toxicity of carbon tetrachloride, free radicals and role of antioxidants. Rev Environ Health 36:279–295

Valadi H, Ekström K, Bossios A, Sjöstrand M, Lee JJ, Lötvall JO (2007) Exosome-mediated transfer of mRNAs and microRNAs is a novel mechanism of genetic exchange between cells. Nat Cell Biol 9:654–659

Valipour J, Taghizadeh F, Esfahani R, Ramesh M, Rastegar T (2024) Role of nuclear factor erythroid 2-related factor 2 (Nrf2) in female and male fertility. Heliyon 10:e29752 van de Lagemaat EE, de Groot L, van den Heuvel E (2019) Vitamin B(12) in relation to oxidative stress: a systematic review. Nutrients 11:482

Vanderhout SM, Rastegar Panah M, Garcia-Bailo B, Grace-Farfaglia P, Samsel K, Dockray J, Jarvi K, El-Sohemy A (2021) Nutrition, genetic variation and male fertility. Transl Androl Urol 10:1410–1431

Vermot A, Petit-Hartlein I, Smith SME, Fieschi F (2021) NADPH oxidases (NOX): an overview from discovery, molecular mechanisms to physiology and pathology. Antioxidants (Basel) 10: 890

Vernet P, Fulton N, Wallace C, Aitken RJ (2001) Analysis of reactive oxygen species generating systems in rat epididymal spermatozoa. Biol Reprod 65:1102–1113

Vernet P, Aitken RJ, Drevet JR (2004) Antioxidant strategies in the epididymis. Mol Cell Endocrinol 216:31–39

Vidarsson H, Westergren R, Heglind M, Blomqvist SR, Breton S, Enerback S (2009) The forkhead transcription factor Foxi1 is a master regulator of vacuolar H-ATPase proton pump subunits in the inner ear, kidney and epididymis. PLoS One 4:e4471

Vijayaraghavan S, Hoskins DD (1990) Changes in the mitochondrial calcium influx and efflux properties are responsible for the decline in sperm calcium during epididymal maturation. Mol Reprod Dev 25:186–194

Vinay L, Belleannee C (2022) EV duty vehicles: features and functions of ciliary extracellular vesicles. Front Genet 13:916233

Wahl B, Reichmann D, Niks D, Krompholz N, Havemeyer A, Clement B, Messerschmidt T, Rothkegel M, Biester H, Hille R et al (2010) Biochemical and spectroscopic characterization of the human mitochondrial amidoxime reducing components hmARC-1 and hmARC-2 suggests the existence of a new molybdenum enzyme family in eukaryotes. J Biol Chem 285: 37847–37859

Wajda A, Lapczuk J, Grabowska M, Slojewski M, Laszczynska M, Urasinska E, Drozdzik M (2016) Nuclear factor E2-related factor-2 (Nrf2) expression and regulation in male reproductive tract. Pharmacol Rep 68:101–108

Waldron KJ, Rutherford JC, Ford D, Robinson NJ (2009) Metalloproteins and metal sensing. Nature 460:823–830

Waldrop GL, Holden HM, St Maurice M (2012) The enzymes of biotin dependent CO(2) metabolism: what structures reveal about their reaction mechanisms. Protein Sci 21:1597–1619

Wallock LM, Tamura T, Mayr CA, Johnston KE, Ames BN, Jacob RA (2001) Low seminal plasma folate concentrations are associated with low sperm density and count in male smokers and nonsmokers. Fertil Steril 75:252–259

Wang D (2012) Redox chemistry of molybdenum in natural waters and its involvement in biological evolution. Front Microbiol 3:427

Wang Z (2023) Vitamin K inhibits ferroptosis via VKORC1L1. Nat Cell Biol 25:1240

Wang J, Wang J, Zhang HR, Shi HJ, Ma D, Zhao HX, Lin B, Li RS (2009) Proteomic analysis of seminal plasma from asthenozoospermia patients reveals proteins that affect oxidative stress responses and semen quality. Asian J Androl 11:484–491

Wang H, Wang Z, Zhou T, Morris D, Chen S, Li M, Wang Y, Zheng H, Fu W, Yan W (2023) Small RNA shuffling between murine sperm and their cytoplasmic droplets during epididymal maturation. Dev Cell 58(9):779–790.e774

Wang X, Fan X, Zhang J, Wang F, Chen J, Wen Y, Wang L, Li T, Li H, Gu H et al (2024a) hnRNPA2B1 represses the disassembly of arsenite-induced stress granules and is essential for male fertility. Cell Rep 43:113769

Wang Y, Li R, Yang R, Zheng D, Zeng L, Lian Y, Zhu Y, Zhao J, Liang X, Li W et al (2024b) Intracytoplasmic sperm injection versus conventional in-vitro fertilisation for couples with infertility with non-severe male factor: a multicentre, open-label, randomised controlled trial. Lancet 403:924–934

Ward WS (2009) Function of sperm chromatin structural elements in fertilization and development. Mol Hum Reprod 16:30–36

Weffort VRS, Lamounier JA (2024) Hidden hunger – a narrative review. J Pediatr 100(Suppl 1): S10–S17
Wei J, Fang T, Shiraiwa M (2022) Effects of acidity on reactive oxygen species formation from secondary organic aerosols. ACS Environ Au 2:336–345
Weigel Munoz M, Cohen DJ, Da Ros VG, Gonzalez SN, Rebagliati Cid A, Sulzyk V, Cuasnicu PS (2024) Physiological and pathological aspects of epididymal sperm maturation. Mol Asp Med 100:101321
Weissgerber P, Kriebs U, Tsvilovskyy V, Olausson J, Kretz O, Stoerger C, Vennekens R, Wissenbach U, Middendorff R, Flockerzi V et al (2011) Male fertility depends on ca(2)+ absorption by TRPV6 in epididymal epithelia. Sci Signal 4:ra27
WHO (2017) Infertility is a global public health issue. WHO
Wilson C, Terman JR, Gonzalez-Billault C, Ahmed G (2016) Actin filaments-A target for redox regulation. Cytoskeleton (Hoboken) 73:577–595
Wiseman H (1993) Vitamin D is a membrane antioxidant. Ability to inhibit iron-dependent lipid peroxidation in liposomes compared to cholesterol, ergosterol and tamoxifen and relevance to anticancer action. FEBS Lett 326:285–288
Wolf P (1967) The nature and significance of platelet products in human plasma. Br J Haematol 13: 269–288
Wong PY, Gong XD, Leung GPH, Cheuk BLY (2002) Formation of the epididymal fluid microenvironment. In: Robaire B, Hinton BT (eds) The epididymis: from molecular to clinical practice A comprehensive survey of the efferent ducts, the epididymis and the vas deferens. Kluwer Academic/Plenum, New York, pp 119–130
Wu S, Zhang R (2020) CD38-expressing macrophages drive age-related NAD(+) decline. Nat Metab 2:1186–1187
Wu KZ, Li K, Galileo DS, Martin-DeLeon PA (2017) Junctional adhesion molecule A: expression in the murine epididymal tract and accessory organs and acquisition by maturing sperm. Mol Hum Reprod 23:132–140
Wu J, Li Z, Wu Y, Cui N (2024) The crosstalk between exosomes and ferroptosis: a review. Cell Death Discovery 10:170
Xiao W, Loscalzo J (2020) Metabolic responses to reductive stress. Antioxid Redox Signal 32: 1330–1347
Xie N, Zhang L, Gao W, Huang C, Huber PE, Zhou X, Li C, Shen G, Zou B (2020) NAD(+) metabolism: pathophysiologic mechanisms and therapeutic potential. Signal Transduct Target Ther 5:227
Xu B, Abdel-Fattah R, Yang L, Crenshaw SA, Black MB, Hinton BT (2011) Testicular lumicrine factors regulate ERK, STAT, and NFKB pathways in the initial segment of the rat epididymis to prevent apoptosis. Biol Reprod 84:1282–1291
Xu H, Yuan SQ, Zheng Z-H, Yan W (2013) The cytoplasmic droplet may be indicative of sperm motility and normal spermiogenesis. Asian J Androl 15(6):799–805
Yahyavi SK, Boisen IM, Cui Z, Jorsal MJ, Kooij I, Holt R, Juul A, Blomberg Jensen M (2024) Calcium and vitamin D homoeostasis in male fertility. Proc Nutr Soc 83:95–108
Yanez-Mo M, Siljander PR, Andreu Z, Zavec AB, Borras FE, Buzas EI, Buzas K, Casal E, Cappello F, Carvalho J et al (2015) Biological properties of extracellular vesicles and their physiological functions. J Extracell Vesicles 4:27066
Yang Y, Wu J, Wang L, Ji G, Dang Y (2024) Copper homeostasis and cuproptosis in health and disease. MedComm (2020) 5:e724
Yeung CH, Breton S, Setiawan I, Xu Y, Lang F, Cooper TG (2004) Increased luminal pH in the epididymis of infertile c-ros knockout mice and the expression of sodium-hydrogen exchangers and vacuolar proton pump H+-ATPase. Mol Reprod Dev 68:159–168
You DG, Oh BH, Nguyen VQ, Lim GT, Um W, Jung JM, Jeon J, Choi JS, Choi YC, Jung YJ et al (2021) Vitamin A-coupled stem cell-derived extracellular vesicles regulate the fibrotic cascade by targeting activated hepatic stellate cells in vivo. J Control Release 336:285–295

Youngson NA, Uddin GM, Das A, Martinez C, Connaughton HS, Whiting S, Yu J, Sinclair DA, Aitken RJ, Morris MJ (2019) Impacts of obesity, maternal obesity and nicotinamide mononucleotide supplementation on sperm quality in mice. Reproduction 158:169–179

Younus AM, Yamanaka T, Shimada M (2024) The protective effects of antioxidants against endogenous and exogenous oxidative stress on bull sperm. In Vitro Cell Dev Biol Anim 60: 969–982

Yuan S, Zheng H, Zheng Z, Yan W (2013) Proteomic analyses reveal a role of cytoplasmic droplets as an energy source during epididymal sperm maturation. PLoS One 8(10):e77466

Yuan X, Duan Y, Xiao Y, Sun K, Qi Y, Zhang Y, Ahmed Z, Moiani D, Yao J, Li H et al (2022) Vitamin E enhances cancer immunotherapy by reinvigorating dendritic cells via targeting checkpoint SHP1. Cancer Discov 12:1742–1759

Zelko IN, Mariani TJ, Folz RJ (2002) Superoxide dismutase multigene family: a comparison of the CuZn-SOD (SOD1), Mn-SOD (SOD2), and EC-SOD (SOD3) gene structures, evolution, and expression. Free Radic Biol Med 33:337–349

Zeng J, Gao W, Tang Y, Wang Y, Liu X, Yin J, Su X, Zhang M, Kang E, Tian Y et al (2023) Hypoxia-sensitive cells trigger NK cell activation via the KLF4-ASH1L-ICAM-1 axis, contributing to impairment in the rat epididymis. Cell Rep 42:113442

Zhai XW, Zhang YL, Qi Q, Bai Y, Chen XL, Jin LJ, Ma XG, Shu RZ, Yang ZJ, Liu FJ (2013) Effects of molybdenum on sperm quality and testis oxidative stress. Syst Biol Reprod Med 59: 251–255

Zhang W, Jiang X, Bao J, Wang Y, Liu H, Tang L (2018) Exosomes in pathogen infections: A bridge to deliver molecules and link functions. Front Immunol 9:90

Zhang Y, Shi J, Rassoulzadegan M, Tuorto F, Chen Q (2019) Sperm RNA code programmes the metabolic health of offspring. Nat Rev Endocrinol 15:489–498

Zhang C, Lin T, Nie G, Hu R, Pi S, Wei Z, Wang C, Xing C, Hu G (2021a) Cadmium and molybdenum co-induce pyroptosis via ROS/PTEN/PI3K/AKT axis in duck renal tubular epithelial cells. Environ Pollut 272:116403

Zhang Y, Ren L, Sun X, Zhang Z, Liu J, Xin Y, Yu J, Jia Y, Sheng J, Hu GF et al (2021b) Angiogenin mediates paternal inflammation-induced metabolic disorders in offspring through sperm tsRNAs. Nat Commun 12:6673

Zhang BL, Zhang ZP, Shi SM, Shi HJ, DeLeon PA, Shum W (2023) Dynamic elementomics of single-cell ICP-MS-derived signals in normal and calcium pump PMCA4-deficient mouse epididymal sperm during capacitation. Metallomics 15:mfad059

Zhang Y, Chen X, Wang M, Zhu Y, Shi W, Li C, Zhang Z, Taniguchi H, Ao P (2024) Nrf1 acts as a highly-conserved Determinon for maintaining robust redox homeostasis in the eco-evo-devo process of life histories. https://doi.org/10.12074/202405.00265V1

Zhang J, Ali MY, Chong HB, Tien PC, Woods J, Noble C, Vornbaumen T, Ordulu Z, Possemato AP, Harry S et al (2025a) Oxidation of retromer complex controls mitochondrial translation. Nature 641:1048

Zhang S, Wang N, Gao Z, Gao J, Wang X, Xie H, Wang CY, Zhang S (2025b) Reductive stress: the key pathway in metabolic disorders induced by overnutrition. J Adv Res. https://doi.org/10.1016/j.jare.2025.01.012

Zhao H, Yu C, He C, Mei C, Liao A, Huang D (2020) The immune characteristics of the epididymis and the immune pathway of the epididymitis caused by different pathogens. Front Immunol 11: 2115

Zhou W, De Iuliis GN, Dun MD, Nixon B (2018) Characteristics of the Epididymal luminal environment responsible for sperm maturation and storage. Front Endocrinol (Lausanne) 9:59

Zhou W, Stanger SJ, Anderson AL, Bernstein IR, De Iuliis GN, McCluskey A, McLaughlin EA, Dun MD, Nixon B (2019) Mechanisms of tethering and cargo transfer during epididymosome-sperm interactions. BMC Biol 17:35

Zmijewski MA (2022) Nongenomic activities of vitamin D. Nutrients 14:5104

Zoroddu MA, Aaseth J, Crisponi G, Medici S, Peana M, Nurchi VM (2019) The essential metals for humans: a brief overview. J Inorg Biochem 195:120–129

Epididymosomes: Composition and Functions for Sperm Maturation

Laura Orama Méar, Pei-Shiue Tsai, Cottrell Tangella Tamessar, John Even Schjenken, and Brett Nixon

Contents

Abstract This article provides an overview of literature pertaining to epididymosome origin, composition and their functional significance. Broadly, epididymosomes are defined as extracellular vesicles that are secreted by the epididymal epithelium and thereafter facilitate intercellular communication within the male reproductive tract. Epididymosomes fulfil this communication role via their

L. O. Méar
Graduate Institute of Veterinary Medicine, School of Veterinary Medicine, National Taiwan University, Taipei, Taiwan

P.-S. Tsai
Graduate Institute of Veterinary Medicine, School of Veterinary Medicine, National Taiwan University, Taipei, Taiwan

Department of Veterinary Medicine, School of Veterinary Medicine, National Taiwan University, Taipei, Taiwan

Research Center for Developmental Biology and Regenerative Medicine, National Taiwan University, Taipei, Taiwan

C. T. Tamessar, J. E. Schjenken, and B. Nixon (✉)
Priority Research Centre for Reproductive Science, School of Environmental and Life Sciences, Discipline of Biological Sciences, The University of Newcastle, Callaghan, NSW, Australia

Hunter Medical Research Institute, Infertility and Reproduction Research Program, New Lambton Heights, NSW, Australia
e-mail: brett.nixon@newcastle.edu.au

Adv Anat Embryol Cell Biol (2025) 241: 89–104
https://doi.org/10.1007/102_2024_7,

Published Online: 24 September 2024

encapsulation and delivery of a diverse macromolecular payload to recipient cells. This complex cargo includes proteins, lipids, and nucleic acids, which are delivered to maturing spermatozoa, thereby influencing their viability and function. Additionally, epididymosomes have been implicated in the post-translational modification of intrinsic sperm proteins, protection of sperm from oxidative stress and immune surveillance, and in the transmission of epigenetic information capable of mediating intergenerational effects. Hence, continued research into the biogenesis, cargo composition, and functional significance of epididymosomes holds promise for advancing male reproductive health and fertility treatments.

Keywords Epididymis, Epididymosomes, Extracellular vesicles, Fertility, Male fertility, Sperm, Sperm maturation

1 An Overview of Sperm Journey Through the Epididymis

Spermatozoa are functionally immature and incapable of fertilization upon release from the testis. They subsequently acquire the capacity for both progressive motility and to participate in oocyte interaction, thereby becoming functional gametes after migrating through the epididymis. This post-spermatogenesis functional transformation is largely reliant on the complex milieu that spermatozoa encounter within the epididymal lumen, which includes a myriad of secretions originating within the lining epididymal epithelium (Brooks 1983; Hinton et al. 1996).

The epididymis itself consists of a single long and convoluted tubule that lies immediately downstream of the efferent duct and terminates in the vas deferens. This continuous tubular structure is histologically separated into four main regions: the initial segment, caput (head), corpus (body), and cauda (tail), each of which presents unique luminal microenvironments. The formation of such distinct microenvironments is coordinated by regional-specific differences in the abundance and activity of six cells types that predominate within the epididymal epithelium (Fig. 1) (Breton et al. 2016; Cornwall 2009; Sullivan et al. 2019). Among these, principal cells are the major epididymal cell type, contributing as much as ~65 to 80% (depending on the segment) of the total cell number along the entire epididymal duct. In accordance with their abundance, principal cells also hold primary responsibility for the bulk of the absorptive and secretive activity of the epididymal epithelium. A striking feature of these cells is that they display distinct region-specific gene expression programs such that neighbouring cells may carry vastly different gene expression patterns (Cornwall 2009; Rinaldi et al. 2020). Importantly in the context of this review, principal cells are widely held to be the main cell type involved in epididymosome production and release to the epididymal lumen (Martin-DeLeon 2015; Sullivan et al. 2007). Halo cells also exist throughout the epididymal epithelium, where they are believed to function as the primary immune cells within the epididymis (Serre and Robaire 1999). Apical cells and narrow cells are primarily located within the

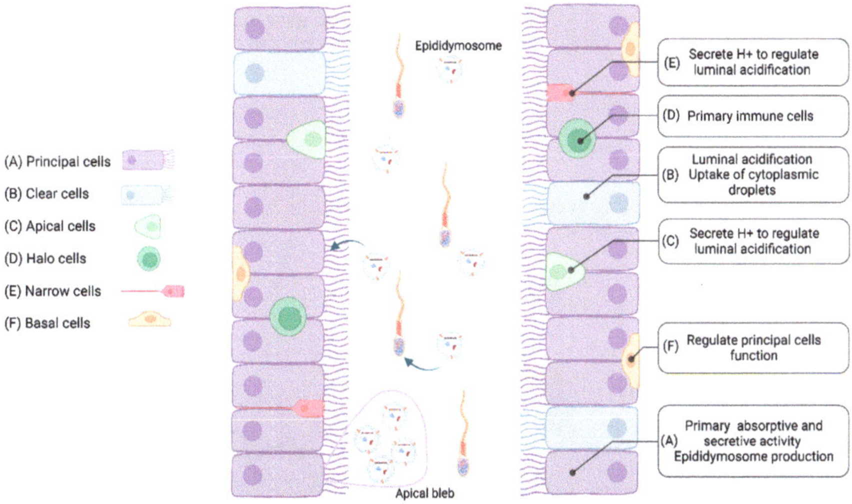

Fig. 1 Key cell types within the epithelial cell layer of the epididymis. Epithelial cells lining the epididymal lumen play an integral role in regional specific differences in the luminal microenvironment. There are 6 main cell types in the epididymal epithelium, including; (A) Principal cells, (B) Clear cells, (C) Apical cells, (D) Halo cells, (E) Narrow cells and (F) Basal cells whose functions are described in more detail above. Amongst these cells, the principal cells hold primary responsibility for the bulk of the absorptive activity and secretion of substances into the epididymal lumen including epididymosomes, which are released from the epididymal epithelium through an apocrine pathway of secretion (apical blebs). Epididymosomes have been implicated in intercellular communication with spermatozoa and downstream epithelial cells

initial segment, where they actively secrete H+ to regulate luminal acidification (Adamali and Hermo 1996; Kujala et al. 2007; Pietrement et al. 2006). In contrast to apical cells, clear cells have been identified throughout the entire epididymal tubule with the exception of the initial segment, and have been implicated not only in luminal acidification but also in the uptake of cytoplasmic droplets released from the surface of maturing sperm cells (Breton et al. 2016; Hermo et al. 2019; Rinaldi et al. 2020). Unlike other epididymal cells, basal cells do not directly contact the luminal compartment but are instead closely associated with the overlying principal cells (Cornwall 2009). Despite this, it has recently been proven that basal cells are capable of extending projections towards the lumen depending on the luminal conditions (Kim et al. 2015; Shum et al. 2008); a phenomenon that likely serves as a dynamic "sensing and bridging system" between epididymal cells. Recent publications also suggest that basal cells may indirectly affect the luminal environment by regulating the functions of the principal cells with which they interact (Leung et al. 2004).

During sperm cells' journey through the epididymis, they experience a continuum of cellular remodelling events, including the binding, attachment, and shedding of molecules in a carefully choreographed sequence. These combined events effectively modify the net sperm surface charge, alter the cell's immunoreactivity, iodination characteristics, capacity for glycolysis, regulate adenylate cyclase

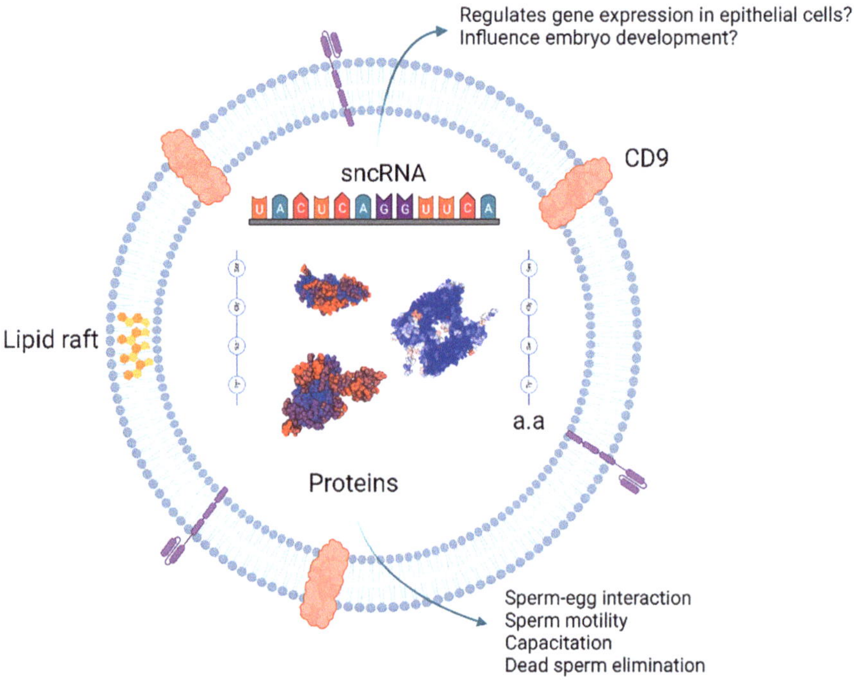

Fig. 2 Schematic depicting epididymosome cargo and putative function

activity/signalling, promote cytoplasmic droplet migration and loss, and profoundly reshape the membrane architecture, including the sperm cells inventory of proteins, phospholipids, and epigenetic cargo (Cornwall 2009; Zhou et al. 2018a). Such changes also manifest in alterations in sperm lectin-binding properties, the extent of acrosome condensation, and the cell's ability to recognise and adhere to the zona pellucida. It is known that many of the components responsible for promoting sperm maturation are secreted by the lining epithelium into the epididymal lumen prior to contacting spermatozoa (Björkgren and Sipilä 2019; Breton et al. 2019; Gervasi and Visconti 2017). Moreover, one of the plausible delivery mechanisms for mediating the bulk exchange of such fertility-modulating cargo is the secretion of extracellular vesicles, termed epididymosomes (Fig. 2) (Sullivan 2015).

2 Defining Features of Epididymosomes

As mentioned above, epididymal epithelial cells are responsible for the production of a multitude of secretory products that are released into the epididymal lumen, whereupon they support sperm maturation, nourishment, and protection (Breton et al. 2016; Cornwall 2009; James et al. 2020; Sullivan et al. 2019). As the dominant form of the secretory pathway within the epididymis, apocrine secretion takes place in the principal cells, where it is initiated by the formation of apical blebs that protrude into the epididymal lumen (Cornwall 2009; Hermo and Jacks 2002; Sullivan 2015). Generally, these blebs contain very few organelles and are devoid of mitochondria, endoplasmic reticulum, and ribosomes; however, they do contain a multitude of smaller membrane-bound extracellular vesicles (also known as epididymosomes) with a diameter of between 50 and 250 nm (Cornwall 2009; Hermo and Jacks 2002; Sullivan 2015). After release and dissolution of the apical blebs, epididymosomes are discharged into the lumen and thereafter form productive interactions with both the luminal sperm cells and non-adjacent epithelial cells leading to the transfer of their complex macromolecular cargo (Belleannée et al. 2013; Frenette et al. 2002; Zhou et al. 2019). Epididymosomes were first identified in the Chinese hamster (Yanagimachi et al. 1985), but have since been described in a number of species, including the human (Frenette et al. 2005; Thimon et al. 2008), mice (Nixon et al. 2019a, b; Rejraji et al. 2006), rat (Martin-DeLeon 2006), domestic cat (Rowlison et al. 2020), sheep (Gatti et al. 2005), horse (Twenter et al. 2020) and bull (Frenette et al. 2002). Structurally, epididymosomes are composed of a phospholipid bilayer enriched in sphingomyelin, cholesterol, transmembrane proteins, and specific extracellular markers (Girouard et al. 2011; Rejraji et al. 2006). Compositionally, epididymosomes comprise a diverse payload of amino acids, proteins, and small non-coding RNAs (sncRNAs) (Fig. 2). Defining the proteomic content of epididymosomes, in particular, has received considerable attention, leading to detailed published inventories in several species, including the mouse (Nixon et al. 2019a, b), bull (Frenette et al. 2010; Frenette et al. 2006; Girouard et al. 2011), ram (Gatti et al. 2005) and human (Thimon et al. 2008). Combined, such studies have led to the curation of >1,500 epididymosome proteins, the majority of which display considerable interspecies conservation.

3 Surface Characteristics of Epididymosomes

In order to study the function of epididymosomes, it has been proven necessary to isolate these vesicles from the epididymal fluid. Depending on the species and hence the size of the epididymis, epididymosomes are usually recovered after making an incision either longitudinally along the epididymis or transversally (Paul et al. 2021). Thereafter, the epididymis is generally incubated in a specific medium to enable aspiration of the luminal contents (Nixon et al. 2019a, b). Fractionation of the epididymosomes can then be achieved by subjecting the whole epididymal fluid to a series of column filtration steps in combination with discontinuous gradient ultracentrifugation. The presence of exosome-enriched proteins, such as CD9 and FLOT1 has been used to verify the presence of epididymosomes in various species (Frenette et al. 2006; Nixon et al. 2019a, b; Paul et al. 2021; Zhou et al. 2018b). Detailed isolation and purification protocols have been elegantly described in many of these studies; however, guidelines produced by the International Society for Extracellular Vesicles (Théry et al. 2018) are driving the extracellular vesicle field towards adopting gold-standard isolation methods. These recommendations include an assessment of EV purity through ultrastructural analysis (e.g. transmission electron microscopy) and an assessment of particle number through light scattering technologies (e.g. nanoparticle tracking analysis). Additionally, epididymosome purity can be measured using a suite of well-established positive (for example, endosome or membrane-binding proteins (TSG101, annexins = ANXA, Rabs = RAB), signal transduction or scaffolding proteins (syntenin), tetraspanins (CD9, CD63, CD81), integrins (ITG) or cell adhesion molecules (CAM), growth factor receptors, heterotrimeric G proteins (GNA), phosphatidylserine-binding MFGE8/lactadherin) and negative (for example, lipoproteins, ribosomal proteins) markers (Théry et al. 2018).

In relation to studies into epididymosome biology, a number of manuscripts have highlighted positive markers detected within these extracellular vesicles. In the mouse, dual labelling of flotillin-1 (FLOT1) and CD9 markers is useful in the detection of epididymosomes (Nixon et al. 2019a, b), while human epididymosomes also present CD9 proteins (Thimon et al. 2008). Surprisingly, bull epididymosomes recovered from the cauda segment showed CD9 positive patterns; however, the protein was undetectable in equivalent populations of caput epididymosomes and only weakly expressed in corpus epididymosomes (Caballero et al. 2013). Moreover, subsequent studies identified two subpopulations of bull epididymosomes, with one being characterized by an abundance of CD9 and capable of transferring proteins to viable maturing spermatozoa and a second population enriched in epididymal sperm binding protein 1 (ELSPBP1) and only capable of interacting with dying or dead epididymal spermatozoa (Caballero et al. 2013; D'Amours et al. 2012). Such findings raise the prospect that different forms of epididymosome may possess discrete functional roles and thus encourage further work to undercover the nuances of the intercellular communication networks they coordinate (Sullivan 2015). Based on these data, it is also apparent that different epididymal segments generate epididymosomes with different lipid profiles (Rejraji et al. 2006). However, epididymosomes collected from distal segments of the epididymis likely represent a mixed population of vesicles containing contributions from upstream epididymal segments (Sullivan and Saez 2013) and potentially also the testes. Apart from variations in epididymosome content, Sullivan et al. also demonstrated that epididymosomes present a complex membrane architecture. Similar to cell membrane structure, epididymosomes possess specialized microdomains known as lipid rafts, which may play a role in the initial docking of sperm and epididymosome membrane as well as facilitate the transfer of cargo proteins (Caballero et al. 2012; Sullivan et al. 2007; Sullivan and Saez 2013).

4 Epididymosome-Sperm Interactions Facilitate the Delivery of Diverse Cargo

During their journey through the epididymis, sperm cells gain functional competency in a process mediated by their contact with epididymal secretions including epididymosomes. These vesicles have gained considerable attention on the basis of their ability to facilitate the selective exchange of fertility-modulating proteins to the recipient sperm cells they encounter (Frenette et al. 2003). Notably, proteomic based studies have led to the identification of several epididymosome proteins that are delivered to sperm and thereafter influence sperm function (Girouard et al. 2011; Nixon et al. 2019a, b; Thimon et al. 2008). Credit for the discovery of sperm: epididymosome exchange of cargo rests with Sullivan and colleagues, whose elegant pioneering work revealed that protein exchange facilitated by epididymosome-sperm interactions is both temperature and pH-dependent and capable of reaching a time-dependent saturation (Frenette et al. 2002). Furthermore, this team also identified that the addition of Zn^{2+} to the co-incubation medium enhanced the interaction between epididymosomes and spermatozoa (Frenette et al. 2002).

Although the exact principles of epididymosome-sperm docking and protein transfer have yet to be fully elucidated, at least three putative mechanisms have been described. The first of these implicates lipid rafts in the direct exchange of proteins, such as P25b, between these respective domains located in the tethered epididymosome and that of the recipient sperm head (Sullivan and Saez 2013). The second mechanism proposes that proteins are trafficked via the formation of transient fusion pores between the epididymosome and underlying sperm plasma membranes (Zhou et al. 2019). During this process, it is envisaged that epididymosome tethering is mediated, at least in part, via interaction between epididymosome ligands (i.e., milk fat globule-EGF factor 8 protein; MFGE8) and complementary αV integrin receptors located in the post-acrosomal domain of the sperm head (Trigg et al. 2021). Thereafter, the engagement of mechanoenzymes such as dynamin 1 mediate the formation and temporary stabilization of fusion pores between the epididymosome and sperm membranes permissive of cargo transfer. An alternative mechanism implicates epididymosomes in the delivery of glycosylphosphatidylinositol-(GPI)-linked proteins to the external lipid layer of the sperm surface (Griffiths et al. 2008; Martin-DeLeon 2015; Patel et al. 2013). By way of example, epididymosomes have been shown to relay transmembrane proteins, including plasma membrane Ca^{2+}-ATPase 4 (PMCA4), to the sperm surface by a fusogenic mechanism facilitated by CD9; a tetraspanin that is responsible for generating fusion–competent sites on membranes (Martin-DeLeon 2015; Patel et al. 2013).

5 The Effects of Epididymosome Trafficked Proteins on Sperm Physiology

It is well-established that the epididymis plays a vital role in the promotion of sperm maturation (Cornwall 2009; Skerrett-Byrne et al. 2022; Sullivan et al. 2019; Sullivan and Mieusset 2016). Proteomic analysis of epididymosome content has greatly helped in elucidating the potential contribution of epididymosome toward this transformation of sperm function. Acting via the mechanisms described above, it is now widely appreciated that epididymosomes facilitate the transfer of a variety of membrane-orientated and intracellular proteins to the recipient sperm population; with a summary of some of the main candidate proteins and their associated functions being provided in Table 1. Among the more well-characterized epididymosome proteins, ZP3R, several belonging ADAM protein family, and GLIPR1L1 have received attention in the context of their conserved putative roles in mediating sperm-oocyte interactions across different species (Girouard et al. 2011; Nixon et al. 2019a, b). However, the function of epididymosomes is not limited to facilitating sperm-oocyte interaction, with several proteins related to capacitation, acrosome reaction, and motility (e.g. CRISP1, AKR1B1) having been found to be present and transferred to maturing spermatozoa by epididymosomes (Table 1) (Gatti et al. 2005; Girouard et al. 2011; Nixon et al. 2019a, b).

In addition to the well-established role of the epididymis in coordinating the functional maturation of spermatozoa, the epididymis also serves to protect sperm cells against oxidative stress and to facilitate their prolonged storage in a viable state. Implicating epididymosomes in these functions, several antioxidant agents have been identified within the encapsulated epididymosome cargo (e.g. glutathione peroxidase, glutathione S-transferase) (Table 1). The most notable endogenous antioxidant detected in epididymosomes is glutathione peroxidase 5 (GPX5) (Taylor et al. 2013); a selenium-independent glutathione peroxidase enzyme that represents a major contributor to the epididymal antioxidant defence system. Specifically, GPX5 is responsible for the enzymatic detoxification of the damaging reactive oxygen species, hydrogen peroxide, into water. In this capacity, GPX5 protects the vulnerable germline against hydrogen peroxide-induced damage to the paternal genome, membrane lipid peroxidation, and protein carbonylation; all of which could otherwise negatively affect sperm quality, fertilization ability, and viability (Takeshima et al. 2021). Other members of the enzymatic and non-enzymatic antioxidant system, for example, peroxiredoxin, glutathione transferase, and ubiquitin, have also been documented as constituents of epididymosomes (Girouard et al. 2011). Taken together, the above-mentioned epididymosomal components maintain a microenvironment that is compatible with sperm maturation, limiting oxidative stress and the accompanying risk it possesses to sperm function. In this respect, it is also noteworthy that epididymosomes bear markers, such as epididymal sperm binding protein 1 (ELSPBP1), that are responsible for the labelling of dead spermatozoa. Although the exact impact of ELSPBP1 on the spermatozoa it is transferred to remains unclear,

Table 1 A summary of epididymosome protein content and their proposed functions in relation to the regulation of sperm physiology

Function	Protein	Species	Reference
Sperm-oocyte interaction	ZP3R, ZPBP	Mouse	(Nixon et al. 2019a, b)
	ADAM2,7		
	CRISP1		
	SPAM1	Human	(Samanta et al. 2018)
	P34H		
	CRISP1	Bull	(Girouard et al. 2011) (Sullivan et al. 2007) (D'Amours et al. 2012)
	ADAM 1, 3A		
	GLIPR1L1		
	SDCBP		
	P25b		
Sperm motility	AKAP12	Mouse	(Nixon et al. 2019a, b)
	CRISP1		
	SORD	Human	(Samanta et al. 2018)
	MIF		
	MIF	Bull	(Girouard et al. 2011) (Sullivan et al. 2007) (D'Amours et al. 2012)
	AK1		
	HK1		
	AKR1B1		
	SDH		
	CRISP1		
	ALDA	Ram	(Gatti et al. 2005)
	SDH		
	AKR1B1		
Capacitation	CRISP1	Human	(Samanta et al. 2018)
	AKR1B1	Bull	(Girouard et al. 2011)
	SDH		
	CRISP1		
	SDH	Ram	(Gatti et al. 2005)
	AKR1B1		
Antioxidant	GPX1,3,6	Mouse	(Nixon et al. 2019a, b)
	GSTM1,2,5,6,7		
	GPX5	Human	(Samanta et al. 2018)
	GPX5	Bull	(Girouard et al. 2011) (Sullivan et al. 2007) (D'Amours et al. 2012)
	GST M1, M4		
	PRDX1,2,4,6		
	GSTM5	Ram	(Gatti et al. 2005)
	SOD		
Marker of dead sperm	ELSPB1	Human	(Samanta et al. 2018)
	HE12		
	UBC	Bull	(Girouard et al. 2011) (Sullivan et al. 2007) (D'Amours et al. 2012)
	ELSPBP1		

the fact that it primarily targets dead sperm cells encourages speculation it may function as a quality control mechanism by marking dead spermatozoa for destruction in the epididymis (D'Amours et al. 2012). Taken together, these data illustrate that epididymosome-mediated protein transfer has the potential to influence different aspects of sperm physiology, including sperm-oocyte binding ability, capacitation, acrosome reaction modulation, motility acquisition, antioxidant redox balance, and dead sperm identification for sperm quality control (Table 1). In doing so, these data clearly show the significant influence epididymosomes exert over sperm fertilization potential.

6 The Effect of Epididymosome RNAs on Sperm Physiology

The intricate coordination of gene expression and the creation of distinct segmental-specific microenvironments in the epididymis requires stringent control mechanisms. It is well-accepted that androgens are the dominant regulator of the epididymal gene expression (Belleannée et al. 2012a, b; Brooks 1983), with some studies even demonstrating that androgen-regulated epididymal proteins can be trafficked by epididymosomes (Suryawanshi et al. 2012). However, androgens alone do not fully account for the regulation of the intricate epididymal luminal microenvironment. Rather, mounting evidence points toward additional tiers of regulation, with several research groups now focusing on the influence that small non-coding RNA species (sncRNAs) exert over epididymal gene expression profiles. sncRNAs exert their regulatory function by targeting messenger RNA (mRNA) for deadenylation, decapping, and degradation, thus inducing translation repression and mRNA silencing at either a transcriptional or post-transcriptional level. In the epididymis, different categories of sncRNAs have been identified, including microRNAs (miRNAs), short interfering RNAs (siRNAs), piwi-interacting RNAs (piRNAs), and transfer RNA (tRNA)-derived RNA fragments (tRFs) (Leung et al. 2016). The complex sncRNA profile in the epididymis has been elucidated in several species, including in the human (Belleannée et al. 2012a, b; Li et al. 2012; Zhang et al. 2010), rat (Zhang et al. 2011), mouse (Nixon et al. 2015a, b) and bull (Belleannée et al. 2013). Together, these studies have revealed that among sncRNA subclasses identified in the mammalian epididymal epithelium, miRNAs rank as the most abundantly expressed and highly conserved (Nixon et al. 2015a, b). The specific sncRNA profile of epididymosomes has also been studied, confirming that proportionally, they contain mainly miRNAs (41–61%) and tRFs (31–40%). Interestingly, the principal cells that are thought to be responsible for epididymosome secretion possess a different sncRNA profile, raising the prospect that sncRNA packaging into epididymosomes might be a selective process (Hutcheon et al. 2017). Furthermore, evidence has shown that epididymosomes are also capable of binding to epithelial cells *in vitro* (Belleannée et al. 2013). It is, therefore, possible that delivery of sncRNA cargo to epididymal epithelial cells by epididymosomes *in vivo* may functionally regulate gene expression patterns in these cells; however, more research is needed to improve understanding of the paracrine regulation exerted by epididymosomes over epididymal epithelium (Björkgren et al. 2015; James et al. 2020; Nixon et al. 2019a, b).

Although sperm cells are rendered both transcriptionally and translationally silent prior to leaving the testes, intriguingly their sncRNA profile undergoes substantial changes coincident with their journey through the epididymis (Hutcheon et al. 2017; Nixon et al. 2015a, b). Specifically, the relative proportions of sncRNA subclasses such as miRNAs, piRNAs, and tRFs are all notably altered during epididymal sperm maturation. Since such alterations to the sperm sncRNA profile are likely driven by extrinsic factors, it stands to reason that epididymosomes likely hold a key contributory role (Trigg et al. 2019). Indeed, *in vitro* studies involving co-incubation of epididymosomes and spermatozoa has proven that epididymosomes are capable of delivering sncRNAs to the sperm cells (Reilly et al. 2016; Sharma et al. 2018). However, mRNA silencing would have little biological relevance for the transcriptionally and translationally inert spermatozoa, encouraging speculation that these epigenetic changes target downstream functions such as the regulation of the early embryonic gene expression landscape (Nixon et al. 2019a, b). While not without some debate, elegant studies by Conine and colleagues have indeed shown that differential loading of sperm-borne sncRNAs can have a profound impact on early embryonic gene expression programs with consequences extending to compromise of embryo development (Conine et al. 2018, 2019, 2020). More recent studies using a corticosterone stress model have implicated epididymosome delivery of sncRNA to sperm to the development of altered offspring phenotypes (Chan et al. 2020). Considering that epididymosomes play a pivotal role in remodelling of the sncRNA profile observed in mature sperm, it appears likely that we are only now beginning to unravel the influence that these extracellular vesicles ultimately have on embryo development and offspring phenotype (James et al. 2020; Yuan et al. 2016).

7 Conclusion and Future Perspectives

Notwithstanding the relatively recent discovery of epididymosomes, considerable progress has been made toward shedding light on their importance in male reproduction. Indeed, despite their nanometer size, the diverse payload of macromolecular cargo that epididymosomes encapsulate has been implicated in not only promoting sperm maturation but also regulating downstream embryo development and offspring health (James et al. 2020; Paul et al. 2021; Rimmer et al. 2021; Trigg et al. 2019). Such findings raise a number of pertinent questions, including the mechanisms regulating epididymosome cargo loading under both normal physiological conditions and in response to paternal stressors, the molecular basis by which epididymosomes differentiate viable from non-viable spermatozoa, whether epididymosome functional roles extend beyond the male reproductive tract to also include conditioning of the peri-conceptual environment within the female reproductive tract, and to the potential of epididymosomes to be harnessed as delivery vectors for the selective trafficking of therapeutic payloads to recipient sperm cells. Ultimately, the field of reproductive biology stands to benefit from continued investigation into the molecular properties and functional roles credited to this unique class of extracellular vesicles.

Compliance with Ethical Standards All authors declare that they have no conflict of interest.
This chapter is a review of previously published accounts, as such, no animal or human studies were performed.

References

Adamali HI, Hermo L (1996) Apical and narrow cells are distinct cell types differing in their structure, distribution, and functions in the adult rat epididymis. J Androl 17(3):208–222

Belleannée C, Calvo E, Thimon V, Cyr DG, Légaré C, Garneau L, Sullivan R (2012a) Role of microRNAs in controlling gene expression in different segments of the human epididymis. PLoS One 7(4):e34996. https://doi.org/10.1371/journal.pone.0034996

Belleannée C, Thimon V, Sullivan R (2012b) Region-specific gene expression in the epididymis. Cell Tissue Res 349(3):717–731. https://doi.org/10.1007/s00441-012-1381-0

Belleannée C, Calvo É, Caballero J, Sullivan R (2013) Epididymosomes convey different repertoires of microRNAs throughout the bovine epididymis. Biol Reprod 89(2):30. https://doi.org/10.1095/biolreprod.113.110486

Björkgren I, Sipilä P (2019) The impact of epididymal proteins on sperm function. Reproduction 158(5):R155–r167. https://doi.org/10.1530/rep-18-0589

Björkgren I, Gylling H, Turunen H, Huhtaniemi I, Strauss L, Poutanen M, Sipilä P (2015) Imbalanced lipid homeostasis in the conditional Dicer1 knockout mouse epididymis causes instability of the sperm membrane. FASEB J 29(2):433–442. https://doi.org/10.1096/fj.14-259382

Breton S, Ruan YC, Park YJ, Kim B (2016) Regulation of epithelial function, differentiation, and remodeling in the epididymis. Asian J Androl 18(1):3–9. https://doi.org/10.4103/1008-682x.165946

Breton S, Nair AV, Battistone MA (2019) Epithelial dynamics in the epididymis: role in the maturation, protection, and storage of spermatozoa. Andrology 7(5):631–643. https://doi.org/10.1111/andr.12632

Brooks DE (1983) Epididymal functions and their hormonal regulation. Aust J Biol Sci 36(3):205–221. https://doi.org/10.1071/bi9830205

Caballero J, Frenette G, D'Amours O, Belleannée C, Lacroix-Pepin N, Robert C, Sullivan R (2012) Bovine sperm raft membrane associated glioma pathogenesis-related 1-like protein 1 (GliPr1L1) is modified during the epididymal transit and is potentially involved in sperm binding to the zona pellucida. J Cell Physiol 227(12):3876–3886. https://doi.org/10.1002/jcp.24099

Caballero JN, Frenette G, Belleannée C, Sullivan R (2013) CD9-positive microvesicles mediate the transfer of molecules to bovine spermatozoa during epididymal maturation. PLoS One 8(6):e65364. https://doi.org/10.1371/journal.pone.0065364

Chan JC, Morgan CP, Adrian Leu N, Shetty A, Cisse YM, Nugent BM, Morrison KE, Jašarević E, Huang W, Kanyuch N, Rodgers AB, Bhanu NV, Berger DS, Garcia BA, Ament S, Kane M, Neill Epperson C, Bale TL (2020) Reproductive tract extracellular vesicles are sufficient to transmit intergenerational stress and program neurodevelopment. Nat Commun 11(1):1499. https://doi.org/10.1038/s41467-020-15305-w

Conine CC, Sun F, Song L, Rivera-Pérez JA, Rando OJ (2018) Small RNAs gained during Epididymal transit of sperm are essential for embryonic development in mice. Dev Cell 46(4):470–480.e473. https://doi.org/10.1016/j.devcel.2018.06.024

Conine CC, Sun F, Song L, Rivera-Pérez JA, Rando OJ (2019) MicroRNAs absent in caput sperm are required for normal embryonic development. Dev Cell 50(1):7–8. https://doi.org/10.1016/j.devcel.2019.06.007

Conine CC, Sun F, Song L, Rivera-Pérez JA, Rando OJ (2020) Sperm head preparation and genetic background affect caput sperm ICSI embryo viability: cauda-enriched miRNAs only essential in specific conditions. Dev Cell 55(6):677–678. https://doi.org/10.1016/j.devcel.2020.11.021

Cornwall GA (2009) New insights into epididymal biology and function. Hum Reprod Update 15(2):213–227. https://doi.org/10.1093/humupd/dmn055

D'Amours O, Frenette G, Bordeleau LJ, Allard N, Leclerc P, Blondin P, Sullivan R (2012) Epididymosomes transfer epididymal sperm binding protein 1 (ELSPBP1) to dead spermatozoa during epididymal transit in bovine. Biol Reprod 87(4):94. https://doi.org/10.1095/biolreprod.112.100990

Frenette G, Lessard C, Sullivan R (2002) Selected proteins of "prostasome-like particles" from epididymal cauda fluid are transferred to epididymal caput spermatozoa in bull. Biol Reprod 67(1):308–313. https://doi.org/10.1095/biolreprod67.1.308

Frenette G, Lessard C, Madore E, Fortier MA, Sullivan R (2003) Aldose reductase and macrophage migration inhibitory factor are associated with epididymosomes and spermatozoa in the bovine epididymis. Biol Reprod 69(5):1586–1592. https://doi.org/10.1095/biolreprod.103.019216

Frenette G, Légaré C, Saez F, Sullivan R (2005) Macrophage migration inhibitory factor in the human epididymis and semen. Mol Hum Reprod 11(8):575–582. https://doi.org/10.1093/molehr/gah197

Frenette G, Girouard J, Sullivan R (2006) Comparison between epididymosomes collected in the intraluminal compartment of the bovine caput and cauda epididymidis. Biol Reprod 75(6):885–890. https://doi.org/10.1095/biolreprod.106.054692

Frenette G, Girouard J, D'Amours O, Allard N, Tessier L, Sullivan R (2010) Characterization of two distinct populations of epididymosomes collected in the intraluminal compartment of the bovine cauda epididymis. Biol Reprod 83(3):473–480. https://doi.org/10.1095/biolreprod.109.082438

Gatti JL, Métayer S, Belghazi M, Dacheux F, Dacheux JL (2005) Identification, proteomic profiling, and origin of ram epididymal fluid exosome-like vesicles. Biol Reprod 72(6):1452–1465. https://doi.org/10.1095/biolreprod.104.036426

Gervasi MG, Visconti PE (2017) Molecular changes and signaling events occurring in spermatozoa during epididymal maturation. Andrology 5(2):204–218. https://doi.org/10.1111/andr.12320

Girouard J, Frenette G, Sullivan R (2011) Comparative proteome and lipid profiles of bovine epididymosomes collected in the intraluminal compartment of the caput and cauda epididymidis. Int J Androl 34(5 Pt 2):e475–e486. https://doi.org/10.1111/j.1365-2605.2011.01203.x

Griffiths GS, Galileo DS, Reese K, Martin-Deleon PA (2008) Investigating the role of murine epididymosomes and uterosomes in GPI-linked protein transfer to sperm using SPAM1 as a model. Mol Reprod Dev 75(11):1627–1636. https://doi.org/10.1002/mrd.20907

Hermo L, Jacks D (2002) Nature's ingenuity: bypassing the classical secretory route via apocrine secretion. Mol Reprod Dev 63(3):394–410. https://doi.org/10.1002/mrd.90023

Hermo L, Oliveira RL, Smith CE, Au CE, Bergeron JJM (2019) Dark side of the epididymis: tails of sperm maturation. Andrology 7(5):566–580. https://doi.org/10.1111/andr.12641

Hinton BT, Palladino MA, Rudolph D, Lan ZJ, Labus JC (1996) The role of the epididymis in the protection of spermatozoa. Curr Top Dev Biol 33:61–102. https://doi.org/10.1016/s0070-2153(08)60337-3

Hutcheon K, McLaughlin EA, Stanger SJ, Bernstein IR, Dun MD, Eamens AL, Nixon B (2017) Analysis of the small non-protein-coding RNA profile of mouse spermatozoa reveals specific enrichment of piRNAs within mature spermatozoa. RNA Biol 14(12):1776–1790. https://doi.org/10.1080/15476286.2017.1356569

James ER, Carrell DT, Aston KI, Jenkins TG, Yeste M, Salas-Huetos A (2020) The role of the epididymis and the contribution of epididymosomes to mammalian reproduction. Int J Mol Sci 21(15). https://doi.org/10.3390/ijms21155377

Kim B, Roy J, Shum WW, Da Silva N, Breton S (2015) Role of testicular luminal factors on Basal cell elongation and proliferation in the mouse epididymis. Biol Reprod 92(1):9. https://doi.org/10.1095/biolreprod.114.123943

Kujala M, Hihnala S, Tienari J, Kaunisto K, Hästbacka J, Holmberg C, Kere J, Höglund P (2007) Expression of ion transport-associated proteins in human efferent and epididymal ducts. Reproduction 133(4):775–784. https://doi.org/10.1530/rep.1.00964

Leung GP, Cheung KH, Leung CT, Tsang MW, Wong PY (2004) Regulation of epididymal principal cell functions by basal cells: role of transient receptor potential (Trp) proteins and cyclooxygenase-1 (COX-1). Mol Cell Endocrinol 216(1–2):5–13. https://doi.org/10.1016/j.mce.2003.10.077

Leung YY, Kuksa PP, Amlie-Wolf A, Valladares O, Ungar LH, Kannan S, Gregory BD, Wang LS (2016) DASHR: database of small human noncoding RNAs. Nucleic Acids Res 44(D1):D216–D222. https://doi.org/10.1093/nar/gkv1188

Li Y, Wang HY, Wan FC, Liu FJ, Liu J, Zhang N, Jin SH, Li JY (2012) Deep sequencing analysis of small non-coding RNAs reveals the diversity of microRNAs and piRNAs in the human epididymis. Gene 497(2):330–335. https://doi.org/10.1016/j.gene.2012.01.038

Martin-DeLeon PA (2006) Epididymal SPAM1 and its impact on sperm function. Mol Cell Endocrinol 250(1–2):114–121. https://doi.org/10.1016/j.mce.2005.12.033

Martin-DeLeon PA (2015) Epididymosomes: transfer of fertility-modulating proteins to the sperm surface. Asian J Androl 17(5):720–725. https://doi.org/10.4103/1008-682x.155538

Nixon B, Stanger SJ, Mihalas BP, Reilly JN, Anderson AL, Dun MD, Tyagi S, Holt JE, McLaughlin EA (2015a) Next generation sequencing analysis reveals segmental patterns of microRNA expression in mouse epididymal epithelial cells. PLoS One 10(8):e0135605. https://doi.org/10.1371/journal.pone.0135605

Nixon B, Stanger SJ, Mihalas BP, Reilly JN, Anderson AL, Tyagi S, Holt JE, McLaughlin EA (2015b) The microRNA signature of mouse spermatozoa is substantially modified during epididymal maturation. Biol Reprod 93(4):91. https://doi.org/10.1095/biolreprod.115.132209

Nixon B, De Iuliis GN, Dun MD, Zhou W, Trigg NA, Eamens AL (2019a) Profiling of epididymal small non-protein-coding RNAs. Andrology 7(5):669–680. https://doi.org/10.1111/andr.12640

Nixon B, De Iuliis GN, Hart HM, Zhou W, Mathe A, Bernstein IR, Anderson AL, Stanger SJ, Skerrett-Byrne DA, Jamaluddin MFB, Almazi JG, Bromfield EG, Larsen MR, Dun MD (2019b) Proteomic profiling of mouse epididymosomes reveals their contributions to post-testicular sperm maturation. Mol Cell Proteomics 18(Suppl 1):S91–s108. https://doi.org/10.1074/mcp.RA118.000946

Patel R, Al-Dossary AA, Stabley DL, Barone C, Galileo DS, Strehler EE, Martin-DeLeon PA (2013) Plasma membrane Ca2+-ATPase 4 in murine epididymis: secretion of splice variants in the luminal fluid and a role in sperm maturation. Biol Reprod 89(1):6. https://doi.org/10.1095/biolreprod.113.108712

Paul N, Talluri TR, Nag P, Kumaresan A (2021) Epididymosomes: a potential male fertility influencer. Andrologia 53(9):e14155. https://doi.org/10.1111/and.14155

Pietrement C, Sun-Wada GH, Silva ND, McKee M, Marshansky V, Brown D, Futai M, Breton S (2006) Distinct expression patterns of different subunit isoforms of the V-ATPase in the rat epididymis. Biol Reprod 74(1):185–194. https://doi.org/10.1095/biolreprod.105.043752

Reilly JN, McLaughlin EA, Stanger SJ, Anderson AL, Hutcheon K, Church K, Mihalas BP, Tyagi S, Holt JE, Eamens AL, Nixon B (2016) Characterisation of mouse epididymosomes reveals a complex profile of microRNAs and a potential mechanism for modification of the sperm epigenome. Sci Rep 6:31794. https://doi.org/10.1038/srep31794

Rejraji H, Sion B, Prensier G, Carreras M, Motta C, Frenoux JM, Vericel E, Grizard G, Vernet P, Drevet JR (2006) Lipid remodeling of murine epididymosomes and spermatozoa during epididymal maturation. Biol Reprod 74(6):1104–1113. https://doi.org/10.1095/biolreprod.105.049304

Rimmer MP, Gregory CD, Mitchell RT (2021) The transformative impact of extracellular vesicles on developing sperm. Reprod Fertil 2(3):R51–r66. https://doi.org/10.1530/raf-20-0076

Rinaldi VD, Donnard E, Gellatly K, Rasmussen M, Kucukural A, Yukselen O, Garber M, Sharma U, Rando OJ (2020) An atlas of cell types in the mouse epididymis and vas deferens. eLife 9. https://doi.org/10.7554/eLife.55474

Rowlison T, Cleland TP, Ottinger MA, Comizzoli P (2020) Novel proteomic profiling of epididymal extracellular vesicles in the domestic cat reveals proteins related to sequential sperm maturation with differences observed between normospermic and teratospermic individuals. Mol Cell Proteomics 19(12):2090–2104. https://doi.org/10.1074/mcp.RA120.002251

Samanta L, Parida R, Dias TR, Agarwal A (2018) The enigmatic seminal plasma: a proteomics insight from ejaculation to fertilization. Reprod Biol Endocrinol 16(1):41. https://doi.org/10.1186/s12958-018-0358-6

Serre V, Robaire B (1999) Distribution of immune cells in the epididymis of the aging Brown Norway rat is segment-specific and related to the luminal content. Biol Reprod 61(3):705–714. https://doi.org/10.1095/biolreprod61.3.705

Sharma U, Sun F, Conine CC, Reichholf B, Kukreja S, Herzog VA, Ameres SL, Rando OJ (2018) Small RNAs are trafficked from the epididymis to developing mammalian sperm. Dev Cell 46(4):481–494.e486. https://doi.org/10.1016/j.devcel.2018.06.023

Shum WW, Da Silva N, McKee M, Smith PJ, Brown D, Breton S (2008) Transepithelial projections from basal cells are luminal sensors in pseudostratified epithelia. Cell 135(6):1108–1117. https://doi.org/10.1016/j.cell.2008.10.020

Skerrett-Byrne DA, Anderson AL, Bromfield EG, Bernstein IR, Mulhall JE, Schjenken JE, Dun MD, Humphrey SJ, Nixon B (2022) Global profiling of the proteomic changes associated with the post-testicular maturation of mouse spermatozoa. Cell Rep 41(7):111655. https://doi.org/10.1016/j.celrep.2022.111655

Sullivan R (2015) Epididymosomes: a heterogeneous population of microvesicles with multiple functions in sperm maturation and storage. Asian J Androl 17(5):726–729. https://doi.org/10.4103/1008-682x.155255

Sullivan R, Mieusset R (2016) The human epididymis: its function in sperm maturation. Hum Reprod Update 22(5):574–587. https://doi.org/10.1093/humupd/dmw015

Sullivan R, Saez F (2013) Epididymosomes, prostasomes, and liposomes: their roles in mammalian male reproductive physiology. Reproduction 146(1):R21–R35. https://doi.org/10.1530/rep-13-0058

Sullivan R, Frenette G, Girouard J (2007) Epididymosomes are involved in the acquisition of new sperm proteins during epididymal transit. Asian J Androl 9(4):483–491. https://doi.org/10.1111/j.1745-7262.2007.00281.x

Sullivan R, Légaré C, Lamontagne-Proulx J, Breton S, Soulet D (2019) Revisiting structure/functions of the human epididymis. Andrology 7(5):748–757. https://doi.org/10.1111/andr.12633

Suryawanshi AR, Khan SA, Joshi CS, Khole VV (2012) Epididymosome-mediated acquisition of MMSDH, an androgen-dependent and developmentally regulated epididymal sperm protein. J Androl 33(5):963–974. https://doi.org/10.2164/jandrol.111.014753

Takeshima T, Usui K, Mori K, Asai T, Yasuda K, Kuroda S, Yumura Y (2021) Oxidative stress and male infertility. Reprod Med Biol 20(1):41–52. https://doi.org/10.1002/rmb2.12353

Taylor A, Robson A, Houghton BC, Jepson CA, Ford WC, Frayne J (2013) Epididymal specific, selenium-independent GPX5 protects cells from oxidative stress-induced lipid peroxidation and DNA mutation. Hum Reprod 28(9):2332–2342. https://doi.org/10.1093/humrep/det237

Théry, C., Witwer, K. W., Aikawa, E., Alcaraz, M. J., Anderson, J. D., Andriantsitohaina, R., Antoniou, A., Arab, T., Archer, F., Atkin-Smith, G. K., Ayre, D. C., Bach, J. M., Bachurski, D., Baharvand, H., Balaj, L., Baldacchino, S., Bauer, N. N., Baxter, A. A., Bebawy, M., . . . Zuba-Surma, E. K. (2018). Minimal information for studies of extracellular vesicles 2018 (MISEV2018): a position statement of the International Society for Extracellular Vesicles and update of the MISEV2014 guidelines. J Extracell Vesicles, 7(1), 1535750. https://doi.org/10.1080/20013078.2018.1535750

Thimon V, Frenette G, Saez F, Thabet M, Sullivan R (2008) Protein composition of human epididymosomes collected during surgical vasectomy reversal: a proteomic and genomic approach. Hum Reprod 23(8):1698–1707. https://doi.org/10.1093/humrep/den181

Trigg NA, Eamens AL, Nixon B (2019) The contribution of epididymosomes to the sperm small RNA profile. Reproduction 157(6):R209–R223. https://doi.org/10.1530/rep-18-0480

Trigg NA, Stanger SJ, Zhou W, Skerrett-Byrne DA, Sipilä P, Dun MD, Eamens AL, De Iuliis GN, Bromfield EG, Roman SD, Nixon B (2021) A novel role for milk fat globule-EGF factor 8 protein (MFGE8) in the mediation of mouse sperm-extracellular vesicle interactions. Proteomics 21(13–14):e2000079. https://doi.org/10.1002/pmic.202000079

Twenter H, Klohonatz K, Davis K, Bass L, Coleman SJ, Bouma GJ, Bruemmer JE (2020) Transfer of MicroRNAs from epididymal epithelium to equine spermatozoa. J Equine Vet 87:102841. https://doi.org/10.1016/j.jevs.2019.102841

Yanagimachi R, Kamiguchi Y, Mikamo K, Suzuki F, Yanagimachi H (1985) Maturation of spermatozoa in the epididymis of the Chinese hamster. Am J Anat 172(4):317–330. https://doi.org/10.1002/aja.1001720406

Yuan S, Schuster A, Tang C, Yu T, Ortogero N, Bao J, Zheng H, Yan W (2016) Sperm-borne miRNAs and endo-siRNAs are important for fertilization and preimplantation embryonic development. Development 143(4):635–647. https://doi.org/10.1242/dev.131755

Zhang J, Liu Q, Zhang W, Li J, Li Z, Tang Z, Li Y, Han C, Hall SH, Zhang Y (2010) Comparative profiling of genes and miRNAs expressed in the newborn, young adult, and aged human epididymides. Acta Biochim Biophys Sin Shanghai 42(2):145–153. https://doi.org/10.1093/abbs/gmp116

Zhang YL, Zhang JS, Zhou YC, Zhao Y, Ni MJ (2011) Identification of microRNAs and application of RNA interference for gene targeting in vivo in the rat epididymis. J Androl 32(6):587–591. https://doi.org/10.2164/jandrol.111.013060

Zhou W, De Iuliis GN, Dun MD, Nixon B (2018a) Characteristics of the epididymal luminal environment responsible for sperm maturation and storage. Front Endocrinol (Lausanne) 9:59. https://doi.org/10.3389/fendo.2018.00059

Zhou W, Sipilä P, De Iuliis GN, Dun MD, Nixon B (2018b) Analysis of epididymal protein synthesis and secretion. J Vis Exp 138. https://doi.org/10.3791/58308

Zhou W, Stanger SJ, Anderson AL, Bernstein IR, De Iuliis GN, McCluskey A, McLaughlin EA, Dun MD, Nixon B (2019) Mechanisms of tethering and cargo transfer during epididymosome-sperm interactions. BMC Biol 17(1):35. https://doi.org/10.1186/s12915-019-0653-5

Seminal Vesicle-Derived Exosomes for the Regulation of Sperm Activity

Wei-Chao Chang, Sheng-Hsiang Li, and Pei-Shiue Tsai

Contents

Abstract The seminal vesicle contributes to a large extent of the semen volume and composition. Removal of seminal vesicle or lack of seminal vesicle proteins leads to decreased fertility. Seminal plasma proteome revealed that seminal fluid contained a wide diversity of proteins. Many of them are known to modulate sperm capacitation

---

Supplementary Information The online version contains supplementary material available at https://doi.org/10.1007/102_2024_6.

W.-C. Chang
Center for Molecular Medicine, China Medical University Hospital, China Medical University, Taichung, Taiwan

S.-H. Li (✉)
Department of Medical Research, MacKay Memorial Hospital, Tamsui, Taiwan

MacKay Junior College of Medicine, Nursing, and Management, Taipei, Taiwan
e-mail: lsh@mmh.org.tw

P.-S. Tsai (✉)
Department of Veterinary Medicine, School of Veterinary Medicine, National Taiwan University, Taipei, Taiwan

Graduate Institute of Veterinary Medicine, School of Veterinary Medicine, National Taiwan University, Taipei, Taiwan

Research Center for Developmental Biology and Regenerative Medicine, National Taiwan University, Taipei, Taiwan
e-mail: psjasontsai@ntu.edu.tw

Adv Anat Embryol Cell Biol (2025) 241: 105–122
https://doi.org/10.1007/102_2024_6,

Published Online: 18 September 2024

and serve as capacitation inhibitors or decapacitation factors. Despite identifying secretory vesicles from the male reproductive tract, such as epididymosomes or prostasomes, isolation, identification, and characterization of seminal vesicle-derived exosomes are still unknown. This chapter aims to review the current understanding of the function of seminal vesicles on sperm physiology and male reproduction and provide ultracentrifugation-based isolation protocols for the isolation of seminal vesicle exosomes. Moreover, via proteomic analysis and functional categorization, a total of 726 proteins IDs were identified in the purified seminal vesicle exosomes fraction. Preliminary data showed seminal vesicle-derived exosomes inhibited sperm capacitation; however, more studies will be needed to reveal other functional involvements of seminal vesicle-derived exosomes on the sperm physiology and, more importantly, how these exosomes interact with sperm membrane to achieve their biological effects.

Keywords Exosomes, Fertility, Membrane modification, Reproductive tract, Seminal vesicle, Spermatozoa

1 Introduction

Testicular sperm cells are considered incompetent for fertilization regardless of their fully developed morphological appearance at the final stage of spermatogenesis. Completing subsequent sequential and coordinated events, such as sperm- genital tract epithelium interactions, sperm membrane surface modifications by components within reproductive tracts, and more importantly, sperm-cumulus oocyte complexes (COCs) interactions, allows successful fusion of gametes and the development of the embryo. After testicular sperm left the seminiferous tubules, loss of the cytoplasmic droplet, activation or inactivation of motility-related signal cascades, alterations on either protein or lipid composition of the sperm plasma membrane all contribute to their function competence for fertilization (Brohi and Huo 2017; Gadella and Boerke 2016; Gadella and Luna 2014; Pitnick et al. 2020). As proactive *de novo* protein translation and membrane recycling by intracellular vesicle transport were ceased in sperm, alterations on sperm membrane surface protein and lipid profiles by components secreted from the reproductive tracts are considered to be the only known mechanism and one of the most prominent prerequisite adjustments for fertilization success (Pitnick et al. 2020; Gadella 2017; Kuo et al. 2016; Tamessar et al. 2021). Post-testicular sperm membrane surface modifications start when sperm cells reach different segments of the epididymis (Gervasi and Visconti 2017). These changes in membrane composition include the addition and removal of proteins, lipids, and even mRNAs to and from the sperm surface. These materials exchange mentioned above is known to result from either a direct sperm-epididymal epithelium interaction (Chen et al. 2021; Shi et al. 2021) or the attachment of epithelium secretory vesicles known as epididymosomes (Tamessar et al. 2021; Barrachina et al. 2022; Baskaran et al. 2020; James et al. 2020; Trigg et al. 2019). After sperm cells leave the epididymis, they mix with seminal fluids upon ejaculation. It is known that seminal fluid maintains semen pH, osmolarity and provides sperm cells with sufficient nutrients for normal motility and survival (Noda and Ikawa 2019; Rodriguez-Martinez et al. 2021). And most of the current available studies focus on the functional roles of seminal vesicle fluid as a whole in the prevention of spontaneous sperm capacitation (Rodriguez-Martinez et al. 2021; Druart and de Graaf 2018; Gonzalez-Cadavid et al. 2014; Maxwell et al. 2007; Szczykutowicz et al. 2019). After sperm cells enter the female reproductive tract, the uterus and oviduct continue to modify the sperm membrane surface via sperm-oviduct interactions or exosomes secreted from the female reproductive tract. These further modifications protect sperm cells from oxidative stress and extend sperm longevity before sperm cells encounter the oocyte (Gadella and Boerke 2016; Pitnick et al. 2020; Al-Dossary and Martin-Deleon 2016; Fujihara et al. 2018; Gonzalez-Brusi et al. 2020; Martin-DeLeon 2016).

2 The Roles of Exosomes on Sperm Membrane Surface Modifications

Exosomes are small secretory vesicles ranging from 30 to 150 nm (with an average vesicle diameter of 100 nm) in size with heterogeneous contents, such as cytosolic or membrane proteins, lipids, nucleic acid, amino acids acid, metabolites, mRNA, and microRNA (Kalluri and LeBleu 2020; Alzhrani et al. 2021). As exosomes are subsets of extracellular vesicles, their biogenesis, contents, functions, and even medical applications vary depending on their origin of secretion (Kalluri and LeBleu 2020; Pegtel and Gould 2019; Hessvik and Llorente 2018). The physiological purposes of exosome secretion are still largely unknown, but the general consensuses are either to remove excessive or unwanted constituents from cells to maintain cellular homeostasis or to actively participate in cell-cell communications using exosome as a cargo delivery system (Barrachina et al. 2022; Kalluri and LeBleu 2020; Pegtel and Gould 2019; van Niel et al. 2018; Stahl and Raposo 2019; He et al. 2021). Recent studies showed that one of the intrinsic properties of exosomes is to regulate complex intracellular signaling pathways; this unique property renders exosomes a potential therapeutic candidate for the treatment of various diseases or medical complications (Qi et al. 2021; Zhang et al. 2019; Zhu et al. 2019). The existence of secretory vesicles and exosomes has been identified from both male and female reproductive tracts, including epididymosomes from the epididymis (Tamessar et al. 2021; Barrachina et al. 2022; James et al. 2020; Trigg et al. 2019; Nixon et al. 2019; Rowlison et al. 2018), prostasomes from the prostate (Vickram et al. 2020; Saez and Sullivan 2016; Aalberts et al. 2014; Ronquist et al. 2006), vaginal exosomes from vaginal-cervical fluids (Fereshteh et al. 2019; Smith and Daniel 2016; Wu et al. 2020a), uterosomes from the uterus (Martin-DeLeon 2016; Franchi et al. 2016; Saribas et al. 2020; Qiao et al. 2018; Kusama et al. 2018) and oviductosomes from the oviduct (Al-Dossary and Martin-Deleon 2016; Lee et al. 2020a, b; Sidrat et al. 2020; Bathala et al. 2018). Functional involvement of exosomes in the facilitation of fertilization processes, such as sperm integrity, membrane composition modification, sperm-oocyte interactions, or even embryo implantation, has also been reviewed in earlier publications (Tamessar et al. 2021; Baskaran et al. 2020; James et al. 2020; Al-Dossary and Martin-Deleon 2016; Mincheva-Nilsson and Baranov 2010; Kowalczyk et al. 2022). These exosomes mainly serve as a cargo transportation carrier that delivers proteins, lipids, mRNA, and microRNA to the subcellular or membranous domains of the sperm cells upon sperm transit in the reproductive tracts. This post-testicular addition of molecules from the reproductive tract not only stabilizes sperm membrane composition to avoid unwanted aggregation but also regulates the timing of capacitation to avoid premature acrosome reaction (Gadella and Boerke 2016; Gadella 2017; Wu et al. 2020b; Wang et al. 2018; Wang et al. 2021; Seligman et al. 2004; Shalgi et al. 1989).

Interestingly, the earlier publications suggested that sperm cells may actively recruit proteasomes upon capacitation. The association of prostasomes to the sperm cells may protect the sperm cells from the attacks of immune cells within the female reproductive tract (Aalberts et al. 2014; Ronquist et al. 2006; Aalberts et al. 2013). Therefore, it is clear that exosomes and their associated components are indispensable for post-testicular sperm maturation and could also serve as novel biomarkers for fertilization capacity or be used as a diagnostic marker for the male fertility (Tamessar et al. 2021; Baskaran et al. 2020; Foot and Kumar 2021; Simon et al. 2018).

3 Seminal Vesicle Secretion on the Modulation of Sperm Physiology

Among different reproductive organs, the seminal vesicle contributes to a large extent of the semen volume and composition. It is known that seminal fluid maintains semen pH, osmolarity and provides sperm cells with sufficient nutrients for normal motility, survival, and the homeostatic anti-oxidation environment to extend sperm longevity (Noda and Ikawa 2019; Rodriguez-Martinez et al. 2021; Hopkins et al. 2017; Barbagallo et al. 2021). Most of the current available studies focus on the functional roles of seminal fluid as a whole, on preventing spontaneous sperm capacitation and unwanted stickiness or aggregation between sperm cells with other components within the reproductive tracts (Gwathmey et al. 2006; Robertson 2007); however, seminal fluid is known as a mixture of secretion from the testis, epididymis, prostate, bulbourethral gland and seminal vesicles (Hopkins et al. 2017). It is evidenced that the removal of seminal vesicle or lack of seminal vesicle proteins (*e.g.*, a heat-labile phospholipid-binding protein, SVS VII, and a carcinoembryonic antigen protein, CEACAM10) led to decrease fertility (Li et al. 2005; Luo et al. 2001). The seminal fluid composition also serves as an indicator for male fertility as the study demonstrated that men with normal or subfertility exhibited very different seminal fluid compositions (Rodriguez-Martinez et al. 2021; Druart and de Graaf 2018; Martins et al. 2020; Yamakawa et al. 2007). Thus, the importance of seminal plasma on fertility is evident; however, the exact mechanism regarding seminal vesicle proteins on the modulation of sperm physiology was not fully understood. Earlier studies using high-resolution mass spectrometry to study seminal plasma proteome revealed that seminal fluid contained a wide diversity of proteins that at least 268 proteins in dog seminal plasma (Aquino-Cortez et al. 2017), 700 in ram seminal plasma (Soleilhavoup et al. 2014), 193 in bovine (Westfalewicz et al. 2017), 563 in porcine (Perez-Patino et al. 2016), 607 in avian species (Atikuzzaman et al. 2017; Labas et al. 2015), and as many as 2000 proteins in human seminal plasma (Rodriguez-Martinez et al. 2021; Gilany et al. 2015; Batruch et al. 2011; Cannarella et al. 2020; Lin et al. 2019). Interestingly, Druart et al. also showed that phylogenetic distances between species might positively correlate to the seminal fluid's proteomic profile, suggesting seminal fluid proteomic profile may reflect the genetic disparity or relatedness among species (Druart and de Graaf 2018). Despite the identification of protein candidates from the seminal fluid in a wide range of mammals, identification, isolation, and functional characterization of exosomes solely from the seminal vesicle secretion is still an ongoing area with limited information.

4 Isolation of Exosomes from the Seminal Vesicles

Earlier studies demonstrated different methods for the isolation and purification of exosomes; these approaches mainly involve in differential ultracentrifugation (Livshits et al. 2015; Linares et al. 2015), ultrafiltration (Cheruvanky et al. 2007; McNamara et al. 2018), aqueous poly-ethylene glycol-based precipitation (Weng et al. 2016), immune-affinity capture (Koliha et al. 2016; Sharma et al. 2018), high throughput microfluidic devices (Hisey et al. 2018; Contreras-Naranjo et al. 2017) or size exclusion chromatography (Sidhom et al. 2020; Rodriguez-Caro et al. 2019; Hong et al. 2016; Driedonks et al. 2019). Among different methodologies, ultracentrifugation, which applies sequential sedimentation steps, is considered the gold standard for exosome isolation and purification. This method separates particles based on their size and density and can yield a relatively larger quantity of exosomes. The ultracentrifugation procedure is performed by separating pellet and supernatant at increasing speeds. Via repeated centrifugation steps, cell debris, cellular components, microvesicles, and exosomes can be separated. The fraction of exosomes is often detected after centrifugation of sample at 100,000–200,000 g for 1–2 h; specific proteins, such as Flotillin-1, Alix, TSG101, CD81, CD63, CD9, ceramide, and syntenin are often used as markers to identify the existence and the purity of the exosomes (Nixon et al. 2019; Wang et al. 2018; Doyle and Wang 2019; Street et al. 2017). However, exosomes of different origins are heterogeneous in population, and in many cases, these so-called "exosome markers" might be absent (Kowal et al. 2016; Sahlen et al. 2010). We developed an ultracentrifugation-based isolation protocol based on the rationality mentioned above to purify mouse seminal vesicle-derived exosomes. As shown in Fig. 1, after seminal vesicle fluid was collected from a direct squeezing of mouse seminal vesicle fluid into a phosphate-

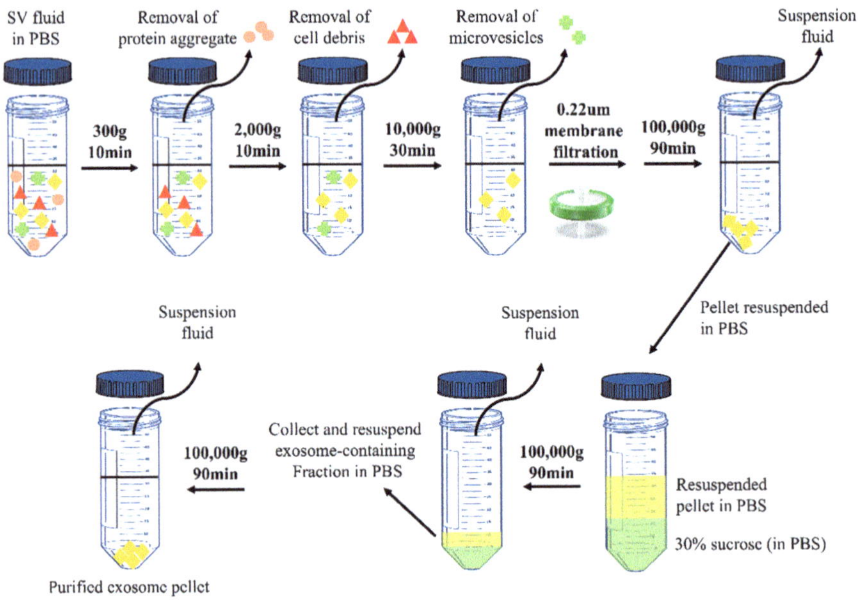

Fig. 1 Workflow for the ultracentrifugation-based mouse seminal vesicle exosome isolation

buffered saline (PBS)-containing tubes, differential centrifugation steps were carried out at 300 g, 2000 g, 10,000 g for 10 min to remove protein aggregation, cell debris, and microvesicles, respectively. A 0.22 μm PVDF membrane was used to filter out potential large extracellular vesicle aggregation caused by the ultracentrifugation (Linares et al. 2015). Exosomes were further concentrated by loading the exosome-containing solution on top of a 30% sucrose layer. After 100,000 g centrifugation for 90 min, the exosome-containing fraction was collected and washed with PBS before a final centrifugation step at 100,000 g for 90 min. The pellet was resuspended in an appropriate quantity of PBS for further characterization.

5 Characterization of Seminal Vesicle-Derived Exosomes

Characterization of exosomes can be based on their ultrastructure, particle size, and surface markers (Alzhrani et al. 2021; Pegtel and Gould 2019; Stahl and Raposo 2019; Doyle and Wang 2019). The ultrastructure of the exosomes from various sources should all be an enclosed membrane bilayer under electron microscopy. Although the known size of the exosomes has an average vesicle diameter of 100 nm, subpopulations of exosomes with different sizes and surface markers have also been described (Kalluri and LeBleu 2020; Pegtel and Gould 2019; Kowal et al. 2016). Large exosomes with the size of 100–160 nm contain a surface

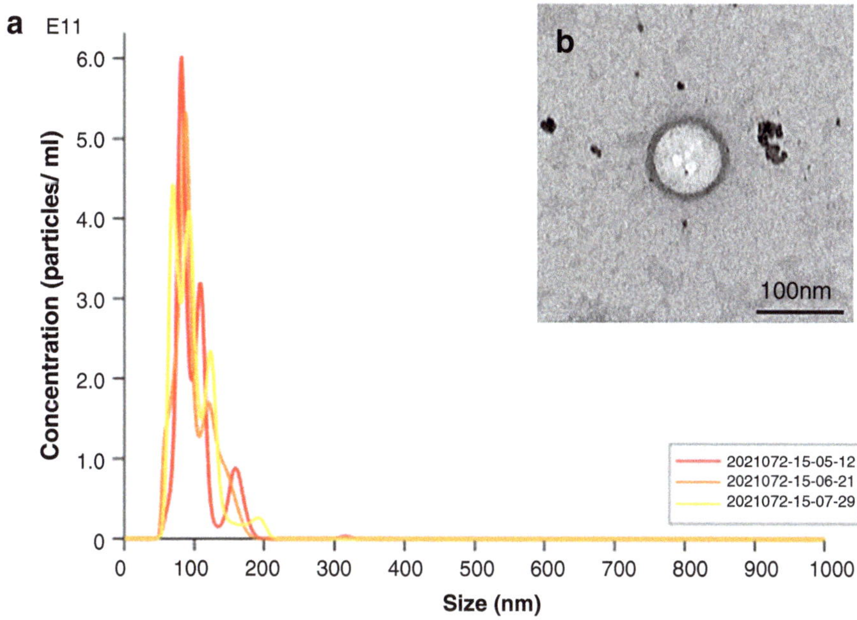

Fig. 2 (**a**) Nanoparticle tracker analysis (NTA) on mouse seminal vesicle exosome particle size. (**b**) The transmission electron micrograph showed a bilamellar membrane vesicle with a size of ~100 nm from the purified seminal vesicle exosome fraction

marker of CD81, and they have been shown to participate in the immune modulation of recipient cells. Medium exosomes with the size from 75 to 100 nm exhibit a surface marker of CD9 and have been shown to initiate the pro-apoptotic signaling pathway of recipient cells. Small exosomes with a particle diameter of 40–75 nm contain a surface marker of CD63 and are considered to activate the pro-survival signaling pathway of recipient cells (Kalluri and LeBleu 2020; Willms et al. 2018). When we subjected purified 100,000 g fraction to nanoparticle tracker analysis (NTA), as indicated in Fig. 2a, isolated vesicles exhibit an average particle size of 83–99 nm (average particle size 89.6 ± 5.9 nm), which fall into the range of defined size for medium exosomes. A transmission electron micrograph also identified the presence of a bilamellar membrane vesicle structure with a diameter of around 100 nm (Fig. 2b). Quantification analysis further demonstrated that using the above-illustrated isolation protocol, a concentration of 8.98×10^9 particles/μg protein can be obtained.

Seminal plasma is known to form a copulatory plug, modulate sperm physiology, and optimize the microenvironment of female reproductive tracts (Noda and Ikawa 2019; Rodriguez-Martinez et al. 2021; Druart and de Graaf 2018; Szczykutowicz et al. 2019; Yamakawa et al. 2007; Soleilhavoup et al. 2014). Although earlier publication suggested that seminal plasma or secretions from accessory glands may not be essential for fertilization but serves to optimize the uterine environment

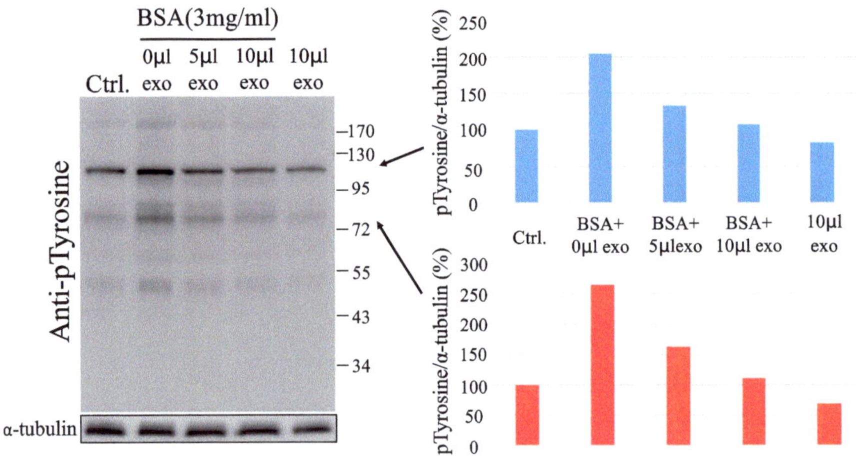

Fig. 3 Inhibitory effects of seminal vesicle-derived exosomes on BSA-induced sperm capacitation

when a low number of spermatozoa are present (Noda and Ikawa 2019; Rodriguez-Martinez et al. 2021); however, most of the current studies support the close associations between seminal vesicles components and sperm functions, such as enhancing sperm membrane stability, sperm motility and sperm survival (Rodriguez-Martinez et al. 2021; Druart and de Graaf 2018; Szczykutowicz et al. 2019; Santiago-Moreno and Blesbois 2020), preventing premature capacitation and acrosome reaction (Wang et al. 2021; Li et al. 2005; Hasan et al. 2021; Li et al. 2018; Pavaneli et al. 2020), and protecting sperm cells from oxidative damages (Barbagallo et al. 2021; Barranco et al. 2021; Ribeiro et al. 2021). Quantitative proteomic analysis showed that the main categories of seminal vesicle proteins are to facilitate vesicle biogenesis, metabolism, membrane adhesion, and remodeling (Druart and de Graaf 2018; Soleilhavoup et al. 2014; Cannarella et al. 2020; Leahy et al. 2020); however, these studies also demonstrated the differences in protein compositions between seminal fluid, sperm membrane surface, seminal vesicle secretion and even between subpopulations of secretory vesicles, (Kalluri and LeBleu 2020; Stahl and Raposo 2019; Cannarella et al. 2020; Doyle and Wang 2019; Kowal et al. 2016; Leahy et al. 2020). In agreement with the known inhibitory function of seminal plasma on sperm capacitation and acrosome reaction, our preliminary data showed that the addition of seminal vesicle-derived exosomes inhibits sperm tyrosine phosphorylation/capacitation (Fig. 3).

Based on proteomic analysis using LC/MS/MS, a total of 726 proteins IDs were identified in the purified seminal vesicle exosomes fraction (Supplementary Table S1). From KEGG enrichment analysis, pathways related to the ribosome, proteasome, and protein processing in the endoplasmic reticulum are significantly enriched (Fig. 4). Further gene ontology (GO) analysis on biological process (BP, Fig. 5a), cellular component (CC, Fig. 5b), and molecular function (MF, Fig. 5c),

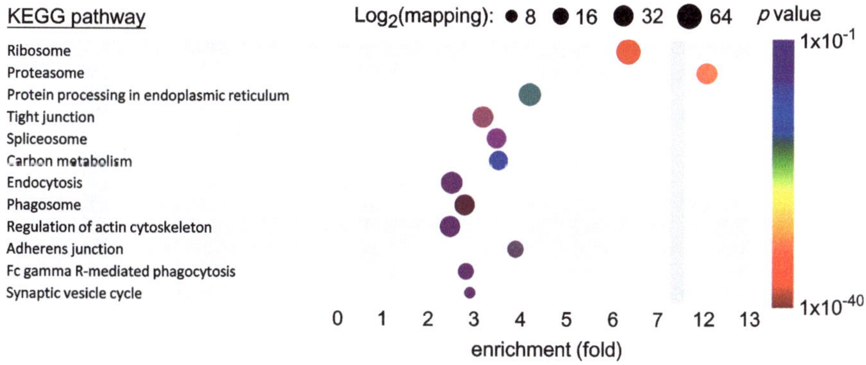

Fig. 4 Scatter plot of KEGG pathway enrichment. The enrichment fold is the ratio of the protein number to the total protein number in a certain pathway. The color and size of the dots represent the range of the p-value and the number of proteins mapped to the indicated pathways, respectively. The most enriched pathways are shown in the figure mapped to the indicated functions or pathways, respectively. The most enriched ones are shown in the figure

indicated that exosome proteins are mostly cytosolic proteins that are responsible for the translational process upon structural constituent of ribosome. Moreover, many proteins identified within the seminal vesicle exosomes are known to exhibit protein, RNA, nucleotide, and actin filament binding processes (Fig. 5). The current proteomic database showed that seminal vesicle secretomes from seminal vesicle secretion (SVS) contain 168 proteins (Smyth et al. 2022); when comparing the existing seminal vesicle secretome database with the seminal vesicle exosome result, only 34 proteins IDs were overlapped. One of these overlapping proteins is a highly enriched seminal vesicle secretory protein IV (SVS4). These distinct protein profiles suggest that the seminal vesicle exosome isolation protocol illustrated in Fig. 1 can efficiently exclude highly enriched seminal vesicle secretory proteins and obtain a relatively pure and distinct exosome population. Earlier studies showed that SVS4 exhibited anti-apoptotic and immunomodulatory effects (Prevete et al. 2010; Morelli et al. 2007); moreover, through the interaction with SVS3, SVS4, similar to SVS2, can serve as a capacitation inhibitor that prevents sperm undergo spontaneous capacitation (Araki et al. 2016). Numerous studies showed that epididymis-derived exosomes (i.e., epididymosomes) could serve as cargo carriers that deliver lipid, protein, mRNA, and specific nucleic acid onto sperm to facilitate the sperm maturation process (Sullivan 2015; Martin-DeLeon 2015; Zhou et al. 2019). Despite the actual function of seminal vesicle-derived exosomes on sperm physiology is still unknown, based on the distinct secretome profiles, it is possible that (1) seminal vesicle exosomes serve as "cargo molecules" like the epididymosomes and deliver specific molecules onto sperm upon ejaculation via direct sperm-exosome interaction, or alternatively, (2) interact with secretory vesicles from other accessory glands (e.g., prostasomes), similar to SVS protein interactions, to control the timing of sperm capacitation and other essential activities. More studies, including the direct

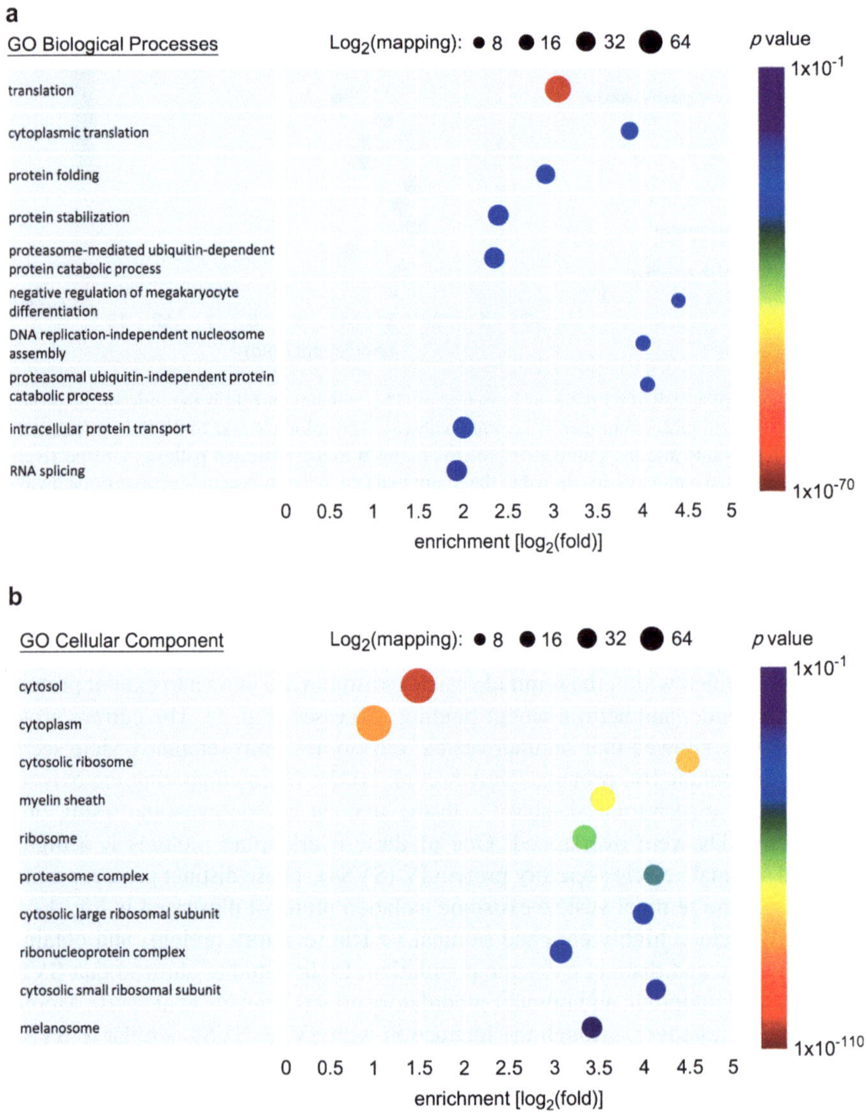

Fig. 5 Scatter plot of the enrichment of GO biological processes (**a**), cellular component (**b**), and molecular function (**c**). The enrichment fold is the ratio of the protein number to the total protein number in a specific function or pathway. The color and size of the dots represent the range of the p-value and the number of proteins mapped to the indicated functions or pathways. Respectively. The most enriched ones are shown in the figure

c

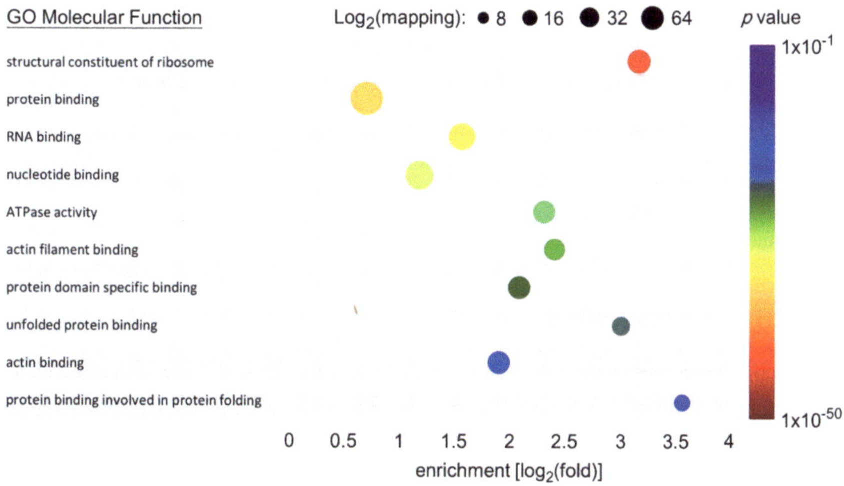

Fig. 5 (continued)

sperm-seminal vesicle exosome co-incubation experiments or genetically modified animals targeting specific seminal vesicle exosome resident proteins, will be needed to elaborate further on the function and the importance of seminal vesicle exosome on sperm physiology and male fertility.

Acknowledgements This study was supported by grants from the Ministry of Science and Technology (Grant #105-2628-B-002-022-MY3 to PS TSAI; #110-2314-B-195-006 to SH LI), Taiwan, and the Ministry of Education (Grant #109L7865 to PS TSAI), Taiwan. This work was also supported by MacKay Memorial Hospital (Grant # MMH 11129 to SH LI), Taipei, Taiwan. The funding bodies had no role in the design of the study and collection, analysis, and interpretation of data.

Compliance with Ethical Standards Ethics approval and sample collections were performed and supervised by the approved veterinarians throughout the study. Material collections from the animals were monitored, and under the guidance, regulation, and approval of the animal welfare committee of National Taiwan University (NTU-104-EL-00081) and Mackay Memorial Hospital (MMH-M-S-110-39).

References

Aalberts M, Sostaric E, Wubbolts R, Wauben MW, Nolte-'t Hoen EN, Gadella BM, Stout TA, Stoorvogel W (2013) Spermatozoa recruit prostasomes in response to capacitation induction. Biochim Biophys Acta 1834:2326–2335. https://doi.org/10.1016/j.bbapap.2012.08.008

Aalberts M, Stout TA, Stoorvogel W (2014) Prostasomes: extracellular vesicles from the prostate. Reproduction 147:R1–R14. https://doi.org/10.1530/REP-13-0358

Al-Dossary AA, Martin-Deleon PA (2016) Role of exosomes in the reproductive tract Oviductosomes mediate interactions of oviductal secretion with gametes/early embryo. Front Biosci (Landmark Ed) 21:1278–1285. https://doi.org/10.2741/4456

Alzhrani GN, Alanazi ST, Alsharif SY, Albalawi AM, Alsharif AA, Abdel-Maksoud MS, Elsherbiny N (2021) Exosomes: isolation, characterization, and biomedical applications. Cell Biol Int 45:1807–1831. https://doi.org/10.1002/cbin.11620

Aquino-Cortez A, Pinheiro BQ, Lima DBC, Silva HVR, Mota-Filho AC, Martins JAM, Rodriguez-Villamil P, Moura AA, Silva LDM (2017) Proteomic characterization of canine seminal plasma. Theriogenology 95:178–186. https://doi.org/10.1016/j.theriogenology.2017.03.016

Araki N, Kawano N, Kang W, Miyado K, Yoshida K, Yoshida M (2016) Seminal vesicle proteins SVS3 and SVS4 facilitate SVS2 effect on sperm capacitation. Reproduction 152:313–321. https://doi.org/10.1530/REP-15-0551

Atikuzzaman M, Sanz L, Pla D, Alvarez-Rodriguez M, Ruber M, Wright D, Calvete JJ, Rodriguez-Martinez H (2017) Selection for higher fertility reflects in the seminal fluid proteome of modern domestic chicken. Comp Biochem Physiol Part D Genomics Proteomics 21:27–40. https://doi.org/10.1016/j.cbd.2016.10.006

Barbagallo F, La Vignera S, Cannarella R, Crafa A, Calogero AE, Condorelli RA (2021) The relationship between seminal fluid Hyperviscosity and oxidative stress: a systematic review. Antioxidants (Basel) 10. https://doi.org/10.3390/antiox10030356

Barrachina F, Battistone MA, Castillo J, Mallofre C, Jodar M, Breton S, Oliva R (2022) Sperm acquire epididymis-derived proteins through epididymosomes. Hum Reprod. https://doi.org/10.1093/humrep/deac015

Barranco I, Rubio CP, Tvarijonaviciute A, Rodriguez-Martinez H, Roca J (2021) Measurement of oxidative stress index in seminal plasma can predict in vivo fertility of liquid-stored porcine artificial insemination semen doses. Antioxidants (Basel):10. https://doi.org/10.3390/antiox10081203

Baskaran S, Panner Selvam MK, Agarwal A (2020) Exosomes of male reproduction. Adv Clin Chem 95:149–163. https://doi.org/10.1016/bs.acc.2019.08.004

Bathala P, Fereshteh Z, Li K, Al-Dossary AA, Galileo DS, Martin-DeLeon PA (2018) Oviductal extracellular vesicles (oviductosomes, OVS) are conserved in humans: murine OVS play a pivotal role in sperm capacitation and fertility. Mol Hum Reprod 24:143–157. https://doi.org/10.1093/molehr/gay003

Batruch I, Lecker I, Kagedan D, Smith CR, Mullen BJ, Grober E, Lo KC, Diamandis EP, Jarvi KA (2011) Proteomic analysis of seminal plasma from normal volunteers and post-vasectomy patients identifies over 2000 proteins and candidate biomarkers of the urogenital system. J Proteome Res 10:941–953. https://doi.org/10.1021/pr100745u

Brohi RD, Huo LJ (2017) Posttranslational modifications in spermatozoa and effects on male fertility and sperm viability. OMICS 21:245–256. https://doi.org/10.1089/omi.2016.0173

Cannarella R, Barbagallo F, Crafa A, La Vignera S, Condorelli RA, Calogero AE (2020) Seminal plasma transcriptome and proteome: towards a molecular approach in the diagnosis of idiopathic male infertility. Int J Mol Sci 21. https://doi.org/10.3390/ijms21197308

Chen H, Alves MBR, Belleannee C (2021) Contribution of epididymal epithelial cell functions to sperm epigenetic changes and the health of progeny. Hum Reprod Update 28:51–66. https://doi.org/10.1093/humupd/dmab029

Cheruvanky A, Zhou H, Pisitkun T, Kopp JB, Knepper MA, Yuen PS, Star RA (2007) Rapid isolation of urinary exosomal biomarkers using a nanomembrane ultrafiltration concentrator. Am J Physiol Renal Physiol 292:F1657–F1661. https://doi.org/10.1152/ajprenal.00434.2006

Contreras-Naranjo JC, Wu HJ, Ugaz VM (2017) Microfluidics for exosome isolation and analysis: enabling liquid biopsy for personalized medicine. Lab Chip 17:3558–3577. https://doi.org/10.1039/c7lc00592j

Doyle LM, Wang MZ (2019) Overview of extracellular vesicles, their origin, composition, purpose, and methods for exosome isolation and analysis. Cells 8. https://doi.org/10.3390/cells8070727

Driedonks TAP, Nijen Twilhaar MK, Nolte-'t Hoen ENM (2019) Technical approaches to reduce interference of fetal calf serum derived RNA in the analysis of extracellular vesicle RNA from cultured cells. J Extracell Vesicles 8:1552059. https://doi.org/10.1080/20013078.2018.1552059
Druart X, de Graaf S (2018) Seminal plasma proteomes and sperm fertility. Anim Reprod Sci 194: 33–40. https://doi.org/10.1016/j.anireprosci.2018.04.061
Fereshteh Z, Bathala P, Galileo DS, Martin-DeLeon PA (2019) Detection of extracellular vesicles in the mouse vaginal fluid: their delivery of sperm proteins that stimulate capacitation and modulate fertility. J Cell Physiol 234:12745–12756. https://doi.org/10.1002/jcp.27894
Foot NJ, Kumar S (2021) The role of extracellular vesicles in sperm function and male fertility. Subcell Biochem 97:483–500. https://doi.org/10.1007/978-3-030-67171-6_19
Franchi A, Cubilla M, Guidobaldi HA, Bravo AA, Giojalas LC (2016) Uterosome-like vesicles prompt human sperm fertilizing capability. Mol Hum Reprod 22:833–841. https://doi.org/10.1093/molehr/gaw066
Fujihara Y, Miyata H, Ikawa M (2018) Factors controlling sperm migration through the oviduct revealed by gene-modified mouse models. Exp Anim 67:91–104. https://doi.org/10.1538/expanim.17-0153
Gadella BM (2017) Reproductive tract modifications of the boar sperm surface. Mol Reprod Dev 84:822–831. https://doi.org/10.1002/mrd.22821
Gadella BM, Boerke A (2016) An update on post-ejaculatory remodeling of the sperm surface before mammalian fertilization. Theriogenology 85:113–124. https://doi.org/10.1016/j.theriogenology.2015.07.018
Gadella BM, Luna C (2014) Cell biology and functional dynamics of the mammalian sperm surface. Theriogenology 81:74–84. https://doi.org/10.1016/j.theriogenology.2013.09.005
Gervasi MG, Visconti PE (2017) Molecular changes and signaling events occurring in spermatozoa during epididymal maturation. Andrology 5:204–218. https://doi.org/10.1111/andr.12320
Gilany K, Minai-Tehrani A, Savadi-Shiraz E, Rezadoost H, Lakpour N (2015) Exploring the human seminal plasma proteome: an unexplored gold mine of biomarker for male infertility and male reproduction disorder. J Reprod Infertil 16:61–71
Gonzalez-Brusi L, Algarra B, Moros-Nicolas C, Izquierdo-Rico MJ, Aviles M, Jimenez-Movilla M (2020) A comparative view on the oviductal environment during the periconception period. Biomol Ther 10. https://doi.org/10.3390/biom10121690
Gonzalez-Cadavid V, Martins JA, Moreno FB, Andrade TS, Santos AC, Monteiro-Moreira AC, Moreira RA, Moura AA (2014) Seminal plasma proteins of adult boars and correlations with sperm parameters. Theriogenology 82:697–707. https://doi.org/10.1016/j.theriogenology.2014.05.024
Gwathmey TM, Ignotz GG, Mueller JL, Manjunath P, Suarez SS (2006) Bovine seminal plasma proteins PDC-109, BSP-A3, and BSP-30-kDa share functional roles in storing sperm in the oviduct. Biol Reprod 75:501–507. https://doi.org/10.1095/biolreprod.106.053306
Hasan MM, Reshi QUA, Lattekivi F, Viil J, Godakumara K, Dissanayake K, Andronowska A, Jaakma U, Fazeli A (2021) Bovine follicular fluid derived extracellular vesicles modulate the viability, capacitation and acrosome reaction of bull spermatozoa. Biology (Basel) 10. https://doi.org/10.3390/biology10111154
He C, Li L, Wang L, Meng W, Hao Y, Zhu G (2021) Exosome-mediated cellular crosstalk within the tumor microenvironment upon irradiation. Cancer Biol Med 18:21–33. https://doi.org/10.20892/j.issn.2095-3941.2020.0150
Hessvik NP, Llorente A (2018) Current knowledge on exosome biogenesis and release. Cell Mol Life Sci 75:193–208. https://doi.org/10.1007/s00018-017-2595-9
Hisey CL, Dorayappan KDP, Cohn DE, Selvendiran K, Hansford DJ (2018) Microfluidic affinity separation chip for selective capture and release of label-free ovarian cancer exosomes. Lab Chip 18:3144–3153. https://doi.org/10.1039/c8lc00834e
Hong CS, Funk S, Muller L, Boyiadzis M, Whiteside TL (2016) Isolation of biologically active and morphologically intact exosomes from plasma of patients with cancer. J Extracell Vesicles 5: 29289. https://doi.org/10.3402/jev.v5.29289

Hopkins BR, Sepil I, Wigby S (2017) Seminal fluid. Curr Biol 27:R404–R405. https://doi.org/10.1016/j.cub.2017.03.063

James ER, Carrell DT, Aston KI, Jenkins TG, Yeste M, Salas-Huetos A (2020) The role of the epididymis and the contribution of epididymosomes to mammalian reproduction. Int J Mol Sci 21. https://doi.org/10.3390/ijms21155377

Kalluri R, LeBleu VS (2020) The biology, function, and biomedical applications of exosomes. Science 367. https://doi.org/10.1126/science.aau6977

Koliha N, Wiencek Y, Heider U, Jungst C, Kladt N, Krauthauser S, Johnston IC, Bosio A, Schauss A, Wild S (2016) A novel multiplex bead-based platform highlights the diversity of extracellular vesicles. J Extracell Vesicles 5:29975. https://doi.org/10.3402/jev.v5.29975

Kowal J, Arras G, Colombo M, Jouve M, Morath JP, Primdal-Bengtson B, Dingli F, Loew D, Tkach M, Thery C (2016) Proteomic comparison defines novel markers to characterize heterogeneous populations of extracellular vesicle subtypes. Proc Natl Acad Sci U S A 113:E968–E977. https://doi.org/10.1073/pnas.1521230113

Kowalczyk A, Wrzecinska M, Czerniawska-Piatkowska E, Kupczynski R (2022) Exosomes – spectacular role in reproduction. Biomed Pharmacother 148:112752. https://doi.org/10.1016/j.biopha.2022.112752

Kuo YW, Li SH, Maeda K, Gadella BM, Tsai PS (2016) Roles of the reproductive tract in modifications of the sperm membrane surface. J Reprod Dev 62:337–343. https://doi.org/10.1262/jrd.2016-028

Kusama K, Nakamura K, Bai R, Nagaoka K, Sakurai T, Imakawa K (2018) Intrauterine exosomes are required for bovine conceptus implantation. Biochem Biophys Res Commun 495:1370–1375. https://doi.org/10.1016/j.bbrc.2017.11.176

Labas V, Grasseau I, Cahier K, Gargaros A, Harichaux G, Teixeira-Gomes AP, Alves S, Bourin M, Gerard N, Blesbois E (2015) Qualitative and quantitative peptidomic and proteomic approaches to phenotyping chicken semen. J Proteomics 112:313–335. https://doi.org/10.1016/j.jprot.2014.07.024

Leahy T, Rickard JP, Pini T, Gadella BM, de Graaf SP (2020) Quantitative proteomic analysis of seminal plasma, sperm membrane proteins, and seminal extracellular vesicles suggests vesicular mechanisms aid in the removal and addition of proteins to the ram sperm membrane. Proteomics 20:e1900289. https://doi.org/10.1002/pmic.201900289

Lee SH, Oh HJ, Kim MJ, Lee BC (2020a) Exosomes derived from oviduct cells mediate the EGFR/MAPK signaling pathway in cumulus cells. J Cell Physiol 235:1386–1404. https://doi.org/10.1002/jcp.29058

Lee SH, Oh HJ, Kim MJ, Lee BC (2020b) Canine oviductal exosomes improve oocyte development via EGFR/MAPK signaling pathway. Reproduction 160:613–625. https://doi.org/10.1530/REP-19-0600

Li SH, Lee RK, Hsiao YL, Chen YH (2005) Demonstration of a glycoprotein derived from the Ceacam10 gene in mouse seminal vesicle secretions. Biol Reprod 73:546–553. https://doi.org/10.1095/biolreprod.105.039651

Li SH, Hwu YM, Lu CH, Lin MH, Yeh LY, Lee RK (2018) Serine protease inhibitor SERPINE2 reversibly modulates murine sperm capacitation. Int J Mol Sci 19. https://doi.org/10.3390/ijms19051520

Lin Y, Liang A, He Y, Li Z, Li Z, Wang G, Sun F (2019) Proteomic analysis of seminal extracellular vesicle proteins involved in asthenozoospermia by iTRAQ. Mol Reprod Dev 86:1094–1105. https://doi.org/10.1002/mrd.23224

Linares R, Tan S, Gounou C, Arraud N, Brisson AR (2015) High-speed centrifugation induces aggregation of extracellular vesicles. J Extracell Vesicles 4:29509. https://doi.org/10.3402/jev.v4.29509

Livshits MA, Khomyakova E, Evtushenko EG, Lazarev VN, Kulemin NA, Semina SE, Generozov EV, Govorun VM (2015) Isolation of exosomes by differential centrifugation: theoretical analysis of a commonly used protocol. Sci Rep 5:17319. https://doi.org/10.1038/srep17319

Luo CW, Lin HJ, Chen YH (2001) A novel heat-labile phospholipid-binding protein, SVS VII, in mouse seminal vesicle as a sperm motility enhancer. J Biol Chem 276:6913–6921. https://doi.org/10.1074/jbc.M006954200
Martin-DeLeon PA (2015) Epididymosomes: transfer of fertility-modulating proteins to the sperm surface. Asian J Androl 17:720–725. https://doi.org/10.4103/1008-682X.155538
Martin-DeLeon PA (2016) Uterosomes: exosomal cargo during the estrus cycle and interaction with sperm. Front Biosci (Schol Ed) 8:115–122. https://doi.org/10.2741/s451
Martins AD, Panner Selvam MK, Agarwal A, Alves MG, Baskaran S (2020) Alterations in seminal plasma proteomic profile in men with primary and secondary infertility. Sci Rep 10:7539. https://doi.org/10.1038/s41598-020-64434-1
Maxwell WM, de Graaf SP, Ghaoui Rel H, Evans G (2007) Seminal plasma effects on sperm handling and female fertility. Soc Reprod Fertil Suppl 64:13–38. https://doi.org/10.5661/rdr-vi-13
McNamara RP, Caro-Vegas CP, Costantini LM, Landis JT, Griffith JD, Damania BA, Dittmer DP (2018) Large-scale, cross-flow based isolation of highly pure and endocytosis-competent extracellular vesicles. J Extracell Vesicles 7:1541396. https://doi.org/10.1080/20013078.2018.1541396
Mincheva-Nilsson L, Baranov V (2010) The role of placental exosomes in reproduction. Am J Reprod Immunol 63:520–533. https://doi.org/10.1111/j.1600-0897.2010.00822.x
Morelli F, Peluso G, Petillo O, Giannattasio A, Filosa S, Motta CM, Tammaro S, Zatterale A, Calzone R, Budillon A et al (2007) The immunomodulatory protein SV-IV protects serum-deprived cells against apoptosis but not against G0/G1 arrest: possible implications for the survival of implanting embryo. J Cell Physiol 212:610–625. https://doi.org/10.1002/jcp.21058
Nixon B, De Iuliis GN, Hart HM, Zhou W, Mathe A, Bernstein IR, Anderson AL, Stanger SJ, Skerrett-Byrne DA, Jamaluddin MFB et al (2019) Proteomic profiling of mouse epididymosomes reveals their contributions to post-testicular sperm maturation. Mol Cell Proteomics 18:S91–S108. https://doi.org/10.1074/mcp.RA118.000946
Noda T, Ikawa M (2019) Physiological function of seminal vesicle secretions on male fecundity. Reprod Med Biol 18:241–246. https://doi.org/10.1002/rmb2.12282
Pavaneli APP, Recuero S, Chaves BR, Garcia-Bonavila E, Llavanera M, Pinart E, Bonet S, De Andrade AFC, Yeste M (2020) The presence of seminal plasma during liquid storage of pig spermatozoa at 17 degrees C modulates their ability to elicit in vitro capacitation and trigger acrosomal exocytosis. Int J Mol Sci 21. https://doi.org/10.3390/ijms21124520
Pegtel DM, Gould SJ (2019) Exosomes. Annu Rev Biochem 88:487–514. https://doi.org/10.1146/annurev-biochem-013118-111902
Perez-Patino C, Barranco I, Parrilla I, Valero ML, Martinez EA, Rodriguez-Martinez H, Roca J (2016) Characterization of the porcine seminal plasma proteome comparing ejaculate portions. J Proteomics 142:15–23. https://doi.org/10.1016/j.jprot.2016.04.026
Pitnick S, Wolfner MF, Dorus S (2020) Post-ejaculatory modifications to sperm (PEMS). Biol Rev Camb Philos Soc 95:365–392. https://doi.org/10.1111/brv.12569
Prevete N, Rossi FW, Triggiani M, Marone G, de Paulis A, Metafora V, De Maria S, Carteni M, Ragone R, Ravagnan G et al (2010) Antiapoptotic seminal vesicle protein IV induces histamine release from human FcepsilonRI+ cells. Int Arch Allergy Immunol 151:318–330. https://doi.org/10.1159/000250440
Qi D, Hou X, Jin C, Chen X, Pan C, Fu H, Song L, Xue J (2021) HNSC exosome-derived MIAT improves cognitive disorders in rats with vascular dementia via the miR-34b-5p/CALB1 axis. Am J Transl Res 13:10075–10093
Qiao F, Ge H, Ma X, Zhang Y, Zuo Z, Wang M, Zhang Y, Wang Y (2018) Bovine uterus-derived exosomes improve developmental competence of somatic cell nuclear transfer embryos. Theriogenology 114:199–205. https://doi.org/10.1016/j.theriogenology.2018.03.027
Ribeiro JC, Braga PC, Martins AD, Silva BM, Alves MG, Oliveira PF (2021) Antioxidants present in reproductive tract fluids and their relevance for fertility. Antioxidants (Basel) 10. https://doi.org/10.3390/antiox10091441

Robertson SA (2007) Seminal fluid signaling in the female reproductive tract: lessons from rodents and pigs. J Anim Sci 85:E36–E44. https://doi.org/10.2527/jas.2006-578
Rodriguez-Caro H, Dragovic R, Shen M, Dombi E, Mounce G, Field K, Meadows J, Turner K, Lunn D, Child T et al (2019) In vitro decidualisation of human endometrial stromal cells is enhanced by seminal fluid extracellular vesicles. J Extracell Vesicles 8:1565262. https://doi.org/10.1080/20013078.2019.1565262
Rodriguez-Martinez H, Martinez EA, Calvete JJ, Pena Vega FJ, Roca J (2021) Seminal plasma: relevant for fertility? Int J Mol Sci 22. https://doi.org/10.3390/ijms22094368
Ronquist KG, Carlsson L, Ronquist G, Nilsson S, Larsson A (2006) Prostasome-derived proteins capable of eliciting an immune response in prostate cancer patients. Int J Cancer 119:847–853. https://doi.org/10.1002/ijc.21895
Rowlison T, Ottinger MA, Comizzoli P (2018) Key factors enhancing sperm fertilizing ability are transferred from the epididymis to the spermatozoa via epididymosomes in the domestic cat model. J Assist Reprod Genet 35:221–228. https://doi.org/10.1007/s10815-017-1083-3
Saez F, Sullivan R (2016) Prostasomes, post-testicular sperm maturation and fertility. Front Biosci (Landmark Ed) 21:1464–1473. https://doi.org/10.2741/4466
Sahlen G, Nilsson O, Larsson A, Carlsson L, Norlen BJ, Ronquist G (2010) Secretions from seminal vesicles lack characteristic markers for prostasomes. Ups J Med Sci 115:107–112. https://doi.org/10.3109/03009730903366067
Santiago-Moreno J, Blesbois E (2020) Functional aspects of seminal plasma in bird reproduction. Int J Mol Sci 21. https://doi.org/10.3390/ijms21165664
Saribas GS, Ozogul C, Tiryaki M, Alpaslan Pinarli F, Hamdemir Kilic S (2020) Effects of uterus derived mesenchymal stem cells and their exosomes on Asherman's syndrome. Acta Histochem 122:151465. https://doi.org/10.1016/j.acthis.2019.151465
Seligman J, Zipser Y, Kosower NS (2004) Tyrosine phosphorylation, thiol status, and protein tyrosine phosphatase in rat epididymal spermatozoa. Biol Reprod 71:1009–1015. https://doi.org/10.1095/biolreprod.104.028035
Shalgi R, Seligman J, Kosower NS (1989) Dynamics of the thiol status of rat spermatozoa during maturation: analysis with the fluorescent labeling agent monobromobimane. Biol Reprod 40: 1037–1045. https://doi.org/10.1095/biolreprod40.5.1037
Sharma P, Ludwig S, Muller L, Hong CS, Kirkwood JM, Ferrone S, Whiteside TL (2018) Immunoaffinity-based isolation of melanoma cell-derived exosomes from plasma of patients with melanoma. J Extracell Vesicles 7:1435138. https://doi.org/10.1080/20013078.2018.1435138
Shi J, Fok KL, Dai P, Qiao F, Zhang M, Liu H, Sang M, Ye M, Liu Y, Zhou Y et al (2021) Spatio-temporal landscape of mouse epididymal cells and specific mitochondria-rich segments defined by large-scale single-cell RNA-seq. Cell Discov 7:34. https://doi.org/10.1038/s41421-021-00260-7
Sidhom K, Obi PO, Saleem A (2020) A review of exosomal isolation methods: is size exclusion chromatography the best option? Int J Mol Sci 21. https://doi.org/10.3390/ijms21186466
Sidrat T, Khan AA, Joo MD, Wei Y, Lee KL, Xu L, Kong IK (2020) Bovine oviduct epithelial cell-derived culture media and exosomes improve mitochondrial health by restoring metabolic flux during pre-implantation development. Int J Mol Sci 21. https://doi.org/10.3390/ijms21207589
Simon C, Greening DW, Bolumar D, Balaguer N, Salamonsen LA, Vilella F (2018) Extracellular vesicles in human reproduction in health and disease. Endocr Rev 39:292–332. https://doi.org/10.1210/er.2017-00229
Smith JA, Daniel R (2016) Human vaginal fluid contains exosomes that have an inhibitory effect on an early step of the HIV-1 life cycle. AIDS 30:2611–2616. https://doi.org/10.1097/QAD.0000000000001236
Smyth SP, Nixon B, Anderson AL, Murray HC, Martin JH, MacDougall LA, Robertson SA, Skerrett-Byrne DA, Schjenken JE (2022) Elucidation of the protein composition of mouse seminal vesicle fluid. Proteomics:e2100227. https://doi.org/10.1002/pmic.202100227

Soleilhavoup C, Tsikis G, Labas V, Harichaux G, Kohnke PL, Dacheux JL, Guerin Y, Gatti JL, de Graaf SP, Druart X (2014) Ram seminal plasma proteome and its impact on liquid preservation of spermatozoa. J Proteomics 109:245–260. https://doi.org/10.1016/j.jprot.2014.07.007

Stahl PD, Raposo G (2019) Extracellular vesicles: exosomes and microvesicles, integrators of homeostasis. Physiology (Bethesda) 34:169–177. https://doi.org/10.1152/physiol.00045.2018

Street JM, Koritzinsky EH, Glispie DM, Yuen PST (2017) Urine exosome isolation and characterization. Methods Mol Biol 1641:413–423. https://doi.org/10.1007/978-1-4939-7172-5_23

Sullivan R (2015) Epididymosomes: a heterogeneous population of microvesicles with multiple functions in sperm maturation and storage. Asian J Androl 17:726–729. https://doi.org/10.4103/1008-682X.155255

Szczykutowicz J, Kaluza A, Kazmierowska-Niemczuk M, Ferens-Sieczkowska M (2019) The potential role of seminal plasma in the fertilization outcomes. Biomed Res Int 2019:5397804. https://doi.org/10.1155/2019/5397804

Tamessar CT, Trigg NA, Nixon B, Skerrett-Byrne DA, Sharkey DJ, Robertson SA, Bromfield EG, Schjenken JE (2021) Roles of male reproductive tract extracellular vesicles in reproduction. Am J Reprod Immunol 85:e13338. https://doi.org/10.1111/aji.13338

Trigg NA, Eamens AL, Nixon B (2019) The contribution of epididymosomes to the sperm small RNA profile. Reproduction 157:R209–R223. https://doi.org/10.1530/REP-18-0480

van Niel G, D'Angelo G, Raposo G (2018) Shedding light on the cell biology of extracellular vesicles. Nat Rev Mol Cell Biol 19:213–228. https://doi.org/10.1038/nrm.2017.125

Vickram AS, Samad HA, Latheef SK, Chakraborty S, Dhama K, Sridharan TB, Sundaram T, Gulothungan G (2020) Human prostasomes an extracellular vesicle – biomarkers for male infertility and prostrate cancer: the journey from identification to current knowledge. Int J Biol Macromol 146:946–958. https://doi.org/10.1016/j.ijbiomac.2019.09.218

Wang TE, Li SH, Minabe S, Anderson AL, Dun MD, Maeda KI, Matsuda F, Chang HW, Nixon B, Tsai PJ (2018) Mouse quiescin sulfhydryl oxidases exhibit distinct epididymal luminal distribution with segment-specific sperm surface associations. Biol Reprod 99:1022–1033. https://doi.org/10.1093/biolre/ioy125

Wang TE, Yeh LY, Kuo-Kuang Lee R, Lu CH, Yang TH, Kuo YW, Joshi R, Tsai PS, Li SH (2021) Secretory mouse quiescin sulfhydryl oxidase 1 aggregates defected human and mouse spermatozoa in vitro and in vivo. iScience 24:103167. https://doi.org/10.1016/j.isci.2021.103167

Weng Y, Sui Z, Shan Y, Hu Y, Chen Y, Zhang L, Zhang Y (2016) Effective isolation of exosomes with polyethylene glycol from cell culture supernatant for in-depth proteome profiling. Analyst 141:4640–4646. https://doi.org/10.1039/c6an00892e

Westfalewicz B, Dietrich MA, Mostek A, Partyka A, Bielas W, Nizanski W, Ciereszko A (2017) Analysis of bull (Bos taurus) seminal vesicle fluid proteome in relation to seminal plasma proteome. J Dairy Sci 100:2282–2298. https://doi.org/10.3168/jds.2016-11866

Willms E, Cabanas C, Mager I, Wood MJA, Vader P (2018) Extracellular vesicle heterogeneity: subpopulations, isolation techniques, and diverse functions in cancer progression. Front Immunol 9:738. https://doi.org/10.3389/fimmu.2018.00738

Wu Y, Wang X, Meng L, Li W, Li C, Li P, Xu S (2020a) Changes of miRNA expression profiles from cervical-vaginal fluid-derived exosomes in response to HPV16 infection. Biomed Res Int 2020:7046894. https://doi.org/10.1155/2020/7046894

Wu L, Ding Y, Han S, Wang Y (2020b) Role of exosomes in the exchange of spermatozoa after leaving the seminiferous tubule: a review. Curr Drug Metab 21:330–338. https://doi.org/10.2174/1389200221666200520091511

Yamakawa K, Yoshida K, Nishikawa H, Kato T, Iwamoto T (2007) Comparative analysis of interindividual variations in the seminal plasma proteome of fertile men with identification of potential markers for azoospermia in infertile patients. J Androl 28:858–865. https://doi.org/10.2164/jandrol.107.002824

Zhang H, Wang L, Li C, Yu Y, Yi Y, Wang J, Chen D (2019) Exosome-induced regulation in inflammatory bowel disease. Front Immunol 10:1464. https://doi.org/10.3389/fimmu.2019.01464

Zhou W, Stanger SJ, Anderson AL, Bernstein IR, De Iuliis GN, McCluskey A, McLaughlin EA, Dun MD, Nixon B (2019) Mechanisms of tethering and cargo transfer during epididymosome-sperm interactions. BMC Biol 17:35. https://doi.org/10.1186/s12915-019-0653-5
Zhu Z, Zhang Y, Zhang Y, Zhang H, Liu W, Zhang N, Zhang X, Zhou G, Wu L, Hua K et al (2019) Exosomes derived from human umbilical cord mesenchymal stem cells accelerate growth of VK2 vaginal epithelial cells through MicroRNAs in vitro. Hum Reprod 34:248–260. https://doi.org/10.1093/humrep/dey344

The Influence of Ovarian-Derived Extracellular Vesicles in Reproduction

Matheus A. Chaves, Juliana G. Ferst, Mariani F. Fiorenza, Franciele F. Vit, and Juliano C. da Silveira

Contents

Abstract In this chapter, we explore the multifaceted roles of extracellular vesicles (EVs) in ovarian biology, focusing on their contributions to folliculogenesis, oocyte competence, corpus luteum function, and immune response regulation. EVs, particularly those derived from follicular fluid (ffEVs), are crucial mediators of cell-to-cell communication within the ovarian follicle, influencing processes such as meiotic progression, stress response, and hormonal regulation. We review preexisting literature, highlighting key findings on the molecular cargo of EVs, such as miRNAs and proteins, and their involvement in regulating the function of the follicle cells. Additionally, the influence of EVs on the immune responses within the ovary was also addressed. Some attention is given to the potential of EVs as non-invasive biomarkers and therapeutic tools, particularly in addressing conditions like premature ovarian insufficiency and polycystic ovary syndrome. By discussing the existing challenges and emerging research, we hope that this chapter will provide a deeper understanding of EVs' therapeutic potential and offer insights or suggestions for advancing assisted reproductive technologies.

J. C. da Silveira (✉)
Laboratory of Molecular Morphophysiology and Development, Department of Veterinary Medicine, School of Animal Science and Food Engineering, University of São Paulo, São Paulo, Brazil
e-mail: julianodasilveira@usp.br

Adv Anat Embryol Cell Biol (2025) 241: 123–146

https://doi.org/10.1007/102_2024_9,

Published Online: 01 January 2025

Keywords Cumulus cells, Extracellular vesicles, Follicular fluid, Granulosa cells, miRNAs, Oocyte maturation

Abbreviations

ADAM17	ADAM metallopeptidase domain 17
AKT	Protein kinase B
BMP15	Bone morphogenetic protein 15
CCs	Cumulus cell
CD11C	Cluster of differentiation 1C
CD163	Cluster of differentiation 63
CD36	Cluster of differentiation 36
CL	Corpus luteum
COC	Cumulus-oocyte complex
DC	Dendric cell
E2	Estrogen
EGFR	Epidermal growth factor receptor
EV	Extracellular vesicle
FF	Follicular fluid
ffEVs	Follicular fluid-derived EVs
FoxO4	Forkhead box protein O4
GCs	Granulosa cell
GDF9	Growth differentiation factor 9
GJ	Gap junction
GRAMD1B	GRAM domain containing 1B
GV	Germinal vesicle
HAS2	Hyaluronan synthase 2
HSP70	Heat shock protein 70
HSP90	Heat shock protein 90
HUCMSC	Human umbilical cord mesenchymal stem cell
IDO	Indoleamine 2,3-dioxygenase 1
IFNG	Interferon gamma
IGFBP2	Insulin-like growth factor binding protein 2
IL10	Interleukin 10
IL1B	Interleukin 1 beta
IL4	Interleukin 4
IL6	Interleukin 6
ILK	Integrin-linked kinase
IVM	In vitro maturation
JAG1	Jagged canonical Notch ligand 1
KISS1	KiSS-1 metastasis suppressor
LH	Luteinizing hormone
MAPK	Mitogen-activated protein kinase

MDK	Midkine
MII	Metaphase II
mTOR	Mechanistic target of rapamycin
MβCD	Methyl-β-cyclodextrin
NF-κB	Nuclear factor kappa B subunit 1
NK	Natural killer cell
NOS2	Nitric oxide synthase 2
NRF2	Nuclear factor erythroid 2–related factor 2
NTF3	Neurotrophin 3
P4	Progesterone
PCOS	Polycystic ovary syndrome
PI3K	Phosphoinositide 3-kinase
POF	Premature ovarian failure
POI	Premature ovarian insufficiency
PTGES2	Prostaglandin E synthase 2
PTX3	Pentraxin 3
ROS	Reactive oxygen species
S100-A9	S100 calcium-binding protein A9
SOD1	Superoxide dismutase type 1
SPTLC2	Serine palmitoyltransferase long chain base subunit 2
Src	Proto-oncogene tyrosine-protein kinase Src
StAR	Steroidogenic acute regulatory protein
STAT3	Signal transducer and activator of transcription 3
TC	Theca cell
TFEB	Transcription factor EB
TGFβ	Transforming growth factor beta
TNF	Tumor necrosis factor
TZP	Transzonal projection
VEGFA	Vascular endothelial growth factor A

1 Introduction

The ovarian follicle is an important microenvironment where the oocyte develops. During the growth of antral follicles, some follicles are selected to become dominant, which can lead, depending on the stage of the estrous cycle, to the ovulation of a mature oocyte (Adams et al. 2008). Concomitant with follicle development, the oocyte grows in this environment and acquires competence to be able to finish maturation, which allows the oocyte to be fertilized and to generate an embryo (Reader et al. 2017; Sirard et al. 2006). At this stage, effective intercellular communication between the ovarian follicle microenvironment and the oocyte is essential to acquire competence. We already know that follicle microenvironment contains EVs secreted by follicular cells (da Silveira et al. 2012) and EVs are important in intercellular communication, acting as paracrine factors. In oocytes, intercellular communication can be mediated by gap junctions (GJ) or transzonal projections (TZPs). EVs were detected near the TZPs, and cells from the follicle environment can internalize them to deliver information, mediating intercellular communication (da Silveira et al. 2012; De Ávila et al. 2020). Moreover, Macaulay et al. (2014) demonstrated that RNA can cross the TZPs and be delivered to the oocyte. Following these discoveries, many studies have highlighted the importance of EVs as mediators of follicular intercellular communication and their crucial role in oocyte development.

During the estrous cycle, the antral follicle is susceptible to hormonal modification which can affect the characteristics of EVs. Since EVs mediate intercellular communication between the follicular microenvironment and the oocyte, changes in this environment can affect the oocyte via EVs. Throughout development, physiological changes occur controlled primarily by gonadotrophins modifying follicular status, which can affect oocyte competence. Many studies have demonstrated that hormones present in the FF are related to oocyte competence and can affect fertility (Matoba et al. 2014; O'Brien et al. 2019). P4 is one of the important hormones that change during the estrous cycle. This hormone also regulates EV secretion and its contents (De Ávila et al. 2020) and can impact oocyte maturation (Peluso et al. 2010; da Silva Rosa et al. 2024a; b). The inhibition of P4 synthesis during IVM decreased cumulus expansion and embryo production rates (Aparicio et al. 2011; O'Shea et al. 2017). In addition, some studies have demonstrated that follicles growing under high P4 concentrations were related to higher oocyte fertility and better cumulus-oocyte interaction in dairy and beef cattle compared with follicles exposed to low P4 concentration (Bisinotto et al. 2010; Denicol et al. 2012; da Silva Rosa et al. 2024a, b). These results demonstrate that P4 is one of the hormones that undergo physiological change during the estrous cycle and can influence molecules such as EVs in this environment, which in turn can affect oocyte competence.

The EV contents in the ovarian follicle microenvironment can also be affected by age and metabolic status. Some studies have demonstrated that EV microRNA (miRNA) contents are distinct in young and old mares as well as in humans (Da Silveira et al. 2015; Diez-Fraile et al. 2014) and they are related to oocyte quality (Zhang et al. 2021). Since the information suggests that EVs present in the follicle microenvironment can possibly be delivered to the oocyte, the EV contents may be related to the decline in oocyte quality with age. The metabolic status can also affect the EV contents. In dairy cows, the metabolic status alters the EVs miRNAs from FF, suggesting that they affect oocyte development competence (Hailay et al. 2019). The lipid profile of EVs from FF is related to oocyte development capacity and some lipids can be associated with the ability of the oocyte to develop to the blastocyst stage (Da Silveira et al. 2021). Thus, EVs can possibly transmit information from the environment to the oocyte, but they can also have a protective effect. Summarily, changes in the follicular environment can be sensed by the cumulus-oocyte complex (COC), potentially altering and compromising its development, with this information being delivered also by EVs. However, EVs

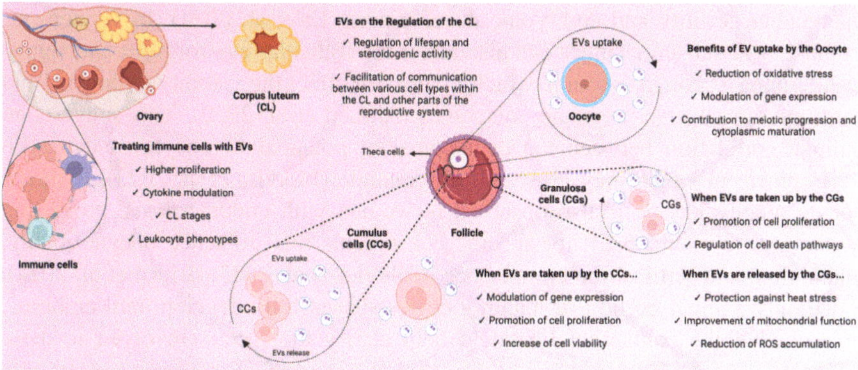

Fig. 1 Summary of the roles of ovarian-derived extracellular vesicles (EVs) in the regulation of ovarian function, including their effects on immune cells, oocytes, and follicular cells (Created in BioRender. Silveira, J. (2024))

also play a protective role and can mitigate the impact of these negative effects through the delivery of beneficial information.

While the role of EVs in safeguarding the oocyte is critical, they are not limited to this function. In fact, their influence extends far beyond the cumulus-oocyte complex, impacting other key cellular players within the ovary. In this chapter, we will explore how ovarian-derived EVs contribute to broader ovarian functions, such as follicular development, luteal regulation, and steroidogenesis. Furthermore, we will examine the immune-modulatory roles of EVs in ovarian tissues, highlighting their influence on immune cell interactions and reproductive health. An illustrative summary is presented in Fig. 1, highlighting the most compelling insights from the current literature. By investigating these diverse roles, we aim to shed light on the overreaching importance of EVs in maintaining ovarian function and their potential applications in improving fertility treatments and diagnosis tools.

2 Role of Follicular Fluid EVs in Oocyte Development

FF plays a crucial role as one of the most important biofluids in animal reproduction. FF is a plasma exudate enriched with a variety of substances secreted by follicular cells such as GCs and CCs, as well as oocytes. Its complex composition includes proteins, hormones, enzymes, polysaccharides, growth factors, cytokines, chemokines, metabolites, anticoagulants, electrolytes, reactive oxygen species, and antioxidants (Moreira et al. 2023). This diverse mixture supports nutrient exchange and facilitates biological signaling between different components of the ovary, which is essential for key processes such as follicle development and oocyte maturation (Brinca et al. 2022; Pocate-Cheriet et al. 2020). As a result, FF plays a fundamental role in determining the quality of the oocytes and influences mammalian fertility and embryonic development (Li et al. 2024). In women, for example, age-related changes can also affect the follicular environment and significantly impact protein function, which in turn can affect the viability of oocytes and contribute to adverse pregnancy outcomes (Ahmed et al. 2020). This suggests an indirect correlation between the developmental potential of oocytes and the FF environment in which they grow and differentiate (Moreira et al. 2023). Reduced oocyte quality has been also observed in women with endometriosis, prompting researchers to examine FF for specific constituents that influence oocyte quality, maturation, and fertilization competence (Collodel et al. 2023; Brinca et al. 2024). Since only a small portion of mature oocytes successfully develop until embryos, investigating the content of FF could help clarify the factors involved in these processes.

Since the composition of FF can be influenced by metabolic diseases, age, and stress factors, a comprehensive characterization of its components may enable the identification of biomarkers associated with oocyte quality and provide insights into the metabolic pathways affected by diseases that impair this quality. Some challenges must be addressed, which include the dynamic changes in FF composition during oocyte growth and development, as well as the fact the fluid's composition varies around each individual oocyte (Montani et al. 2019; Poulsen et al. 2019). It is in this context that EVs can emerge as valuable tools for research in the follicle environment.

EVs are present in the FF of various mammals, including humans, bovines, felines, pigs, and horses (Harris et al. 2020). Studies on ffEVs have shown that these structures play critical roles in oocyte maturation by influencing key mechanisms such as meiotic progression, cytoplasmic maturation, and protection against cellular stress through various pathways (De Ávila & Da Silveira, 2020; Uzbekova et al. 2020). Additionally, molecules such as mRNAs and proteins, released from surrounding follicular cells into the FF via EVs are being studied as non-invasive molecular markers of oocyte competence. In this context, the presence of miRNAs in EVs isolated from FF was first observed by da Silveira et al. (2012) in the FF of mares. Since then, numerous studies have supported the link between the content of EVs, particularly miRNAs, and key pathways of interest in reproduction. It has been also reported that the majority of miRNAs in FF are encapsulated within ffEVs (Sohel et al. 2013). In this regard, ffEVs may regulate the physiological functions of recipient cells by transferring these miRNAs, targeting specific genes (Wang et al. 2021a). For instance, miRNAs present in ffEVs from various species, including humans, cattle, and horses, have been predicted to target and regulate key signaling pathways (Gabryś et al. 2024). These pathways, including the epidermal growth factor receptor (EGFR) pathway, the transforming growth factor-beta (TGFβ), the mitogen-activated protein kinase (MAPK) pathway, and the Wnt signaling pathway, play crucial roles in follicular development, oocyte maturation, and ovulation, highlighting the potential influence of ffEVs on reproductive processes, especially oocyte development (Gabryś et al. 2024).

The follicle environment changes during development and the concentration of ffEVs decreases as the follicle size increases (Navakanitworakul et al. 2016). For example, ffEVs isolated from small follicles in cattle ovaries have been found to promote cell proliferation, whereas EVs from medium and large follicles, which represent later stages of follicular growth, contain miRNAs linked to genes involved in the inflammatory response. This suggests that the content and function of ffEVs change throughout folliculogenesis (Navakanitworakul et al. 2016; Nejabati et al. 2023). Similar findings were observed in the FF of antral follicles from sexually mature gilts, where the concentration of EVs also varied depending on the stage of follicle development (Grzesiak et al. 2020). A similar pattern was also recently observed in buffalo FF (Capra et al. 2022). In this study, the authors confirmed that seasonal changes also influence the miRNA content of ffEVs isolated from both antral and preovulatory follicles. In antral follicles, for example, ffEVs showed dysregulation of miRNAs potentially involved in IL6 expression, while ffEVs from preovulatory follicles were enriched with miRNAs targeting genes related to focal adhesion assembly and cell-substrate junction organization (Capra et al. 2022).

Benedetti et al. (2024) demonstrated that ffEVs from individual cattle follicles contain varying amounts of miRNAs, including bta-miR-34c, which is associated with oocyte competence and the potential to develop into a blastocyst. In their study, the addition of miR-34c mimics during IVM resulted in a higher number of cells in blastocysts, whereas the use of a miR-34c inhibitor led to the opposite effect. Martinez et al. (2018) also reported very promising results when they analyzed miRNAs from ffEVs in oocytes of women whose eggs were either successfully fertilized or not. The study found that ffEVs from the fertilized group contained more miRNAs associated with improved day-3 embryo quality. In another study conducted by the same research group, authors found that increases in the body mass index in women can similarly impact the expression of 18 EV-associated miRNAs in FF. These alterations were linked to developmental pathways of follicles and oocytes, which could potentially lead to reduced fertility (Martinez et al. 2019). In turn, Battaglia et al. (2020) demonstrated that the number of small EVs increases in the FF of older women, primarily due to enhanced release from somatic cells. This increase was then linked to the activation of the TP53 signaling pathways. Through bioinformatics, the authors identified six miRNAs that were deregulated in the FF of older women and discussed that these changes may simultaneously impact oocyte maturation, stress responses, and vesicle secretion. Moreover, Hasan et al. (2020) observed changes in the gene expression of bovine oviductal cells following supplementation with ffEVs, as revealed by transcriptome analysis using mRNA sequencing. The authors compared the effects of supplementing pure FF and isolated ffEVs in cell culture and found different outcomes depending on the duration of incubation. They attributed this difference to the short lifespan of ffEVs, with notable effects observed only during the early stages of co-incubation.

Alternatively, some current studies are also investigating the effects of adding ffEVs to culture media to assess their impact on gamete quality and maturation rates. In Sysoeva et al. (2021), for example, ffEVs were incubated with human sperm, leading to a significant increase in the number of motile spermatozoa, as well as improvements in motility and hyperactivation. The authors attributed these effects to the binding of ffEVs to the acrosomal region of the sperm head, which likely triggered capacitation and the acrosome reaction. This enhancement was also linked to the potential transfer of miRNAs, proteins, and other bioactive molecules essential for energy metabolism from ffEVs to sperm. These findings could interestingly aid in the development of new assisted reproduction techniques for patients with infertility due to impaired sperm motility. On the other hand, Dalanezi et al. (2019) found that EVs from exogenous FF exposed to thermal stress conditions modulated oocyte gene expression differently during IVM of cattle oocytes. The authors observed that ffEVs from cows kept in thermoneutral conditions had reduced expression of genes related to oocyte quality. Gabryś et al. (2022) demonstrated that small ffEVs can enhance the media used for IVM of mare oocytes. Using confocal microscopy, the authors observed that COCs were more likely to reach the metaphase II stage after supplementation with ffEVs in a two-stage culture system. In contrast, the use of pure FF led to changes after 24 h, including an increase in PTGES2, a key enzyme in prostaglandin biosynthesis. Gad et al. (2022) found that supplementing maturation media with ffEVs containing low-quality oocytes influenced the cytoplasmic maturation of oocytes by altering mitochondrial distribution patterns. The authors attributed this effect to the distinct miRNA cargo associated with oocyte quality.

The supplementation with EVs from FF also contributes to embryo development. Asaadi et al. (2021) found that EVs isolated from in vivo-derived FF had a positive effect on embryo development in cattle when ffEVs were supplemented during the first 18 h of maturation. Altogether, these findings suggest that ffEVs have the potential to act as supplements for IVM media to better mimic the in vivo maturation process. However, the effectiveness of this approach may vary depending on the specific biological system being studied.

3 Effects of EVs on Granulosa Cells

Granulosa cells are specialized cells that surround the developing cumulus-oocyte complex within the ovarian follicles, playing a crucial role in oocyte maturation, hormone production, and follicle development. Studies investigating the effects of EVs on GC culture typically focus on changes in gene expression following EV uptake. This effective internalization has been demonstrated by several authors, reflecting a form of intercellular communication within the follicular environment (Da Silveira et al. 2017; Hung et al. 2015; Ying et al. 2023). The mechanisms of EV internalization into GCs have been studied throughout the years, with evidence showing that it primarily occurs through energy-dependent endocytosis (Wang et al. 2024). Among the various pathways, clathrin-mediated endocytosis has been linked to EV uptake (Edure et al. 2024), as well as lipid raft-mediated endocytosis, which is heavily dependent on cholesterol (Gurunathan et al. 2022). These studies have used inhibitors such as heparin, chlorpromazine (for clathrin-mediated pathways), and MβCD (for lipid-rafts pathways). Additionally, Wortmannin and LY294002 have also been employed as inhibitors to explore EV uptake, as this process is also believed to be associated with the PI3K pathway. Recently, Wang et al. (2024) demonstrated that high-density EVs uptake by bovine GCs is time-dependent, relies on tyrosine kinase activity (in addition to PI3K), and involves other pathways beyond clathrin and lipid rafts, including caveolin-mediated endocytosis and micropinocytosis. More recent studies have begun to explore not only the mechanisms of EV uptake but also specific functional outcomes, such as cell proliferation and autophagy, that EVs induce in GCs.

The first study demonstrating the internalization of ffEVs in GCs was conducted by da Silveira et al. (2012), using equine FF. The authors labeled the EVs with PKH67 and observed their internalization into GCs via fluorescence microscopy, both in vitro and in vivo. This study sparked discussions about the role of EVs in mediating intercellular communication in mammalian cells. Moreover, bovine ffEVs from small follicles significantly promoted GC proliferation, an effect that was not observed to the same extent when using EVs from large antral follicles (Hung et al. 2017). The authors demonstrated the EV-induced cell proliferation by showing its prevention using inhibitors of the Src, PI3K, and MAPK signaling pathways. Moreover, high-density EVs were shown to deliver miR-128-3p into GCs, where they activated autophagy and inhibited apoptosis (Ying et al. 2024). In this study, autophagy was confirmed through the activation of targets such as *FoxO4* and *TFEB*. Interestingly, Andrade et al. (2017) observed that miRNAs present in bovine FF were also found in EVs released by bovine GCs into a culture medium, highlighting the extensive communication among follicular cells. These findings on the role of ffEVs in promoting GCs proliferation and regulating cell death pathways underscore the therapeutic potential of EVs in reproductive biology.

In addition to studies supplementing GCs culture media with EVs isolated from different biofluids, some research has focused on investigating the EVs released by GCs themselves into their culture media. These studies often aim to identify miRNAs within EVs that could serve as biomarkers for stressed or toxic environmental conditions. Menjivar et al. (2023) demonstrated that GC-derived EVs play a crucial regulatory role in protecting against heat stress during the IVM of bovine oocytes. The authors observed a reduction in ROS accumulation, an improvement in mitochondrial function, and alterations in the expression of stress-associated genes, ultimately enhancing oocyte survival. The expression profile of EV-associated miRNAs also distinctly separated GCs and their corresponding EVs based on stress exposure (Gebremedhn et al. 2020). The study identified bta-miR-1246 as a potential cellular and extracellular marker of heat stress in GCs, a finding also observed in Holstein Frisian cows (Zheng et al. 2014). Interestingly, EVs released from stressed GCs exhibited a protective effect against subsequent stress by reducing ROS accumulation. Furthermore, EVs from heat-stressed GCs were enriched with HSP70, HSP90, and SOD1, compared to those from untreated cells. On the other hand, GCs-derived EVs contained higher proportions of antioxidants and other defense-related substances in response to oxidative stress compared to normal conditions (Saeed-Zidane et al. 2017). The survival of these cells under oxidative stress was shown to depend on their ability to activate NRF2 response mechanisms. Rodosthenous et al. (2019) demonstrated that the levels of specific EV-enriched miRNAs in the conditioned media of primary GCs vary in response to exposure to supra-physiological levels of bisphenol A. The authors suggested that miRNAs might play a key role as a biological mechanism involved in bisphenol A-induced toxicity within the ovarian follicle. However, they emphasized the need for further studies to validate the role of GC-derived miRNAs in the link between BPA exposure and female infertility. These results demonstrated that EVs are internalized by GCs and regulate important functions in these cells, contributing to follicle development.

4 Effects of EVs on Cumulus Cells (CCs)

Cumulus cells are specialized CGs surrounding the oocyte within the ovarian follicle and play a critical role in nourishing and supporting oocyte development. These cells form a vital component of the COC, which is essential for oocyte maturation and the acquisition of key competencies such as fertilization capability and progression to the blastocyst stage (Zhou et al. 2016). Cumulus expansion not only promotes oocyte maturation but also facilitates follicular rupture during ovulation. Conversely, impaired cumulus expansion leads to reduced ovulation rates and subsequent female infertility (Chen et al. 1993; Gabryś et al. 2024). The role of EVs in COC maturation is currently under investigation. Specific miRNAs and proteins in EVs are known to regulate pathways critical for oocyte development, including cumulus expansion and oocyte quality (Hung et al. 2015, 2017).

Studies have demonstrated that ffEVs can modulate gene expression in cultured CCs (Da Silveira et al. 2017; Da Silveira et al. 2024a, b; Hung et al. 2015). Notably, ffEVs derived from small ovarian follicles, which are a rich source of EVs, have been shown to be readily internalized by CCs (Hung et al. 2015). These ffEVs also exhibit a greater capacity to promote CC proliferation compared to those from medium or large follicles (Hung et al. 2015, 2017; Gabryś et al. 2024). Interestingly, the findings by Hung et al. (2017) challenge the prevailing understanding, as COCs in larger follicles are typically the ones that respond to the LH surge, triggering cumulus expansion. The authors hypothesize that the EVs from small follicles may not be taken up by COCs until after the LH surge, a mechanism that might differ from the behavior of EVs from large follicles, which may fail to promote cumulus expansion. Another possible explanation suggests the presence of inhibitory factors within large follicles that are absent in small antral follicles, which could account for the differential effects on cumulus expansion. In addition, differential proteins and miRNAs have been detected in both ffEVs and CCs from mature and non-mature COCs, highlighting their potential role in the acquisition of oocyte competence (Walter et al. 2020; Uzbekova et al. 2020).

A more focused investigation into the role of ffEVs during the IVM of bovine oocytes was conducted by De Ávila & Da Silveira (2020). The study revealed that CCs begin internalizing EVs as early as 1 h after the start of IVM. Furthermore, the RNA profile of CCs differed depending on whether they were supplemented with ffEVs derived from follicles with low or high P4 levels. Notably, ffEVs from follicles with elevated P4 levels influenced the small EVs, leading to the downregulation of the Notch signaling pathway. In contrast, CCs supplemented with ffEVs from the low-P4 group exhibited upregulation of genes involved in biological processes related to follicular and oocyte development, as well as immune and inflammatory responses, which are essential for CC expansion. Additionally, these EVs might repress RNA translation in CCs, which could explain the observed increase in upregulated genes and associated biological processes. Furthermore, ffEVs derived from cows exposed to heat stress, when added to COC maturation media, resulted in a relatively higher expression of the *IGFBP2, BMP15, GDF9, HAS2,* and *STAT3* in CCs compared to those co-cultured with EVs derived from the FF of cows maintained under thermoneutral conditions. This also highlights the potential of EVs to modulate oocyte gene expression during IVM (Dalanezi et al. 2019). The benefits of adding ffEVs during oocyte maturation and embryo development have also been demonstrated in a heat stress model, where elevated culture temperatures were induced to mimic stress conditions, showing the ability of ffEVs to counteract the negative effects of heat stress (Rodrigues et al. 2019). This protective effect may be attributed to the regulatory molecules present in FF, which can shield adjacent cells and oocytes from heat stress (Tesfaye et al. 2020). In the study by Menjivar et al. (2023), the supplementation of COC maturation media with EVs derived from bovine CGs appeared to alter CC physiology, particularly influencing oocyte physiology by affecting ROS accumulation, mitochondrial activity, and gene expression profiles following heat stress. Notably, the addition of these EVs supported cumulus expansion, both in the presence and absence of heat stress, as evidenced by the increased expression of cumulus expansion-associated genes, including *PTGS2*, *EGFR*, and *PTX3*.

Beyond the role of ffEVs in CC physiology and oocyte maturation, EVs from various other biofluids have also been explored for their capacity to modulate CC function and oocyte competence. Javadi et al. (2022) demonstrated that EVs derived from mouse blood plasma contain components that promote CC maturation and proliferation, accelerating CC expansion. In their study, these EVs upregulated the expression of *Has2* and *Ptgs2*, genes associated with cumulus expansion, through multiple pathways that modulate COC expansion. Similarly, Lee et al. (2020) investigated the effects of EVs from oviduct cell culture media, finding that these EVs improved physiological conditions in CCs treated with GW4869 (an EV formation inhibitor) and Gefitinib (an EGFR inhibitor). The improvements included enhanced cell viability, increased proliferation rates, and reduced ROS levels, primarily mediated through the EGFR/MAPK signaling pathway. Moreover, Abumaghaid et al. (2022) found that EVs derived from rabbit epididymal fluid significantly enhanced oocyte IVM and CC outcomes. This effect was likely linked to the presence of neuropeptides (KISS1, MDK, and NTF3), ADAM metallopeptidase domain 17 (ADAM17), and VEGFA. The authors proposed a novel communication method between male reproductive tract cells and COCs involving the shuttle transfer of mRNA transcripts, offering new insights into the factors influencing the female reproductive tract.

As previously discussed with CGs, CCs can also secrete EVs in culture. Fiorentino et al. (2024) observed EV release from a feeder layer of CCs during the in vitro transition from the GV to the MII stage in mice. The EVs were categorized into two distinct groups based on their content, which correlated with the presence or absence of nucleolus-surrounded oocytes. A broad range of miRNAs were differentially expressed in these two groups, with 7 miRNAs identified as potentially playing a direct role in the acquisition of oocyte competence.

In summary, the diverse studies exploring the impact of EVs, whether from FF, other biofluids, or even secreted by CCs themselves, underscore the critical role of EVs in modulating CC physiology, cumulus expansion, and oocyte competence. These findings highlight the complex and multifaceted nature of EV-mediated communication in reproductive biology, offering promising avenues for improving oocyte maturation protocols and understanding female infertility regulation.

5 Immune Cells and EVs in Ovarian Function

The ovary comprises a diverse subset of cells that work together to support its function. During ovulation, inflammation occurs as part of normal physiological processes (Bukovský et al. 1977; Yang et al. 2019). Immune cells migrate through ovarian structures, contributing to these processes (Yang et al. 2019). The ovarian immune milieu includes macrophages, DCs, NK, granulocytes, and T and B lymphocytes (Abdulrahman and Fair 2019; Bulmer 1964; Enien et al. 1998; Fainaru et al. 2010; Lei et al. 1991; Penny et al. 1999; Smith et al. 2005; Spanel-Borowski et al. 1997; Wu et al. 2004).

One of the first reports identified macrophages in degenerated follicles and the CL (Bulmer 1964). Before ovulation, DCs are present in the ovary, and their numbers increase slowly within the follicle and CL depending on the hormonal environment near ovulation (Wu et al. 2004). Immune cells, including monocytes, are also found within the follicular environment, with a reduction in polymorphonuclear cells observed (Smith et al. 2005). Interestingly, the immune cell population in the follicle differs from that in peripheral blood, potentially indicating an infiltration of immune cells, characteristic of an inflammatory process (Smith et al. 2005). Moreover, the immune cell phenotype varies based on reproductive status. For example, women who retrieved more oocytes during oocyte pick-up procedures had a higher presence of CD56 + CD16-NK cells compared to poor responders (Fainaru et al. 2010).

The corpus hemorrhagicum, originating from the collapsed post-ovulatory follicle, is a heterogeneous structure composed of steroidogenic, endothelial, fibroblast, and immune cells, derived from CGs and TCs of the ruptured follicle (Abdulrahman and Fair 2019; Lei et al. 1991; Penny et al. 1999; Spanel-Borowski et al. 1997). Studies across species have identified eosinophils, neutrophils, and macrophages in the CL (Abdulrahman and Fair 2019; Jiemtaweeboon et al. 2011; Lawler et al. 1999; Penny et al. 1999; Reibiger and Spanel-Borowski 2000; Townson et al. 2002). These cells are recruited by a variety of factors, including cytokines and hormones (Arici et al. 1997; Care et al. 2013; Duncan et al. 2012; Hehnke et al. 1994; Korzekwa et al. 2008; Lawler et al. 1999; Neuvians et al. 2004; Oakley et al. 2010; Okuda et al. 1999; Paavola 1979; Reynolds and Redmer 1999; Townson et al. 2002; Townson and Liptak 2003; Wada et al. 1999). Additionally, EVs play a role in regulating immune cell migration and function in the ovary.

EVs derived from different luteal stages have been used to track T lymphocytes, with the majority of T cells internalizing these EVs. EVs from midcycle and regressing luteal cells induced T cell proliferation in a dose-dependent manner. Midcycle luteal cell-derived EVs upregulated *TNF* mRNA, while those from regressing luteal cells downregulated *IL4, IL10, TNF*, and *IFNG*. Moreover, EVs from both midcycle and regressing luteal cells influenced macrophage phenotypes. Midcycle-derived EVs increased *NOS2, IL10*, and *TGF* expression, decreased *CD163* expression, and had no impact on *TNF, IL1B, IDO*, or *CD36* expression. In contrast, regressing luteal cell-derived EVs increased *NOS2, CD163*, and *CD36* expression, with no changes in *TNF, Il1B, IDO, IL10*, or *TGFB* expression (Branham 2015). CD11B was present in all samples, but CD11C expression was higher in midcycle samples than in regressing ones.

Around ovulation, localized inflammation leads to macrophage infiltration into the ovulatory follicle, where they can polarize into M1 or M2 phenotypes (Xiao et al. 2022). EVs from M1 and M2 macrophages contain specific miRNAs (miR-107 and miR-99a-50, respectively), that activate the PI3K/mTOR signaling pathway, which is critical for ovulation (Xiao et al. 2022). Furthermore, in aged mice, the intravenous administration of EVs derived from M2 macrophages improved ovarian function and alleviated inflammation within the ovary (Xiao et al. 2022). Altogether, these results demonstrate the relationship between EVs and immune cells in the ovary.

6 EVs in Corpus Luteum (CL) Regulation

The CL is a temporary endocrine structure in the ovary that plays a crucial role in regulating the menstrual cycle and supporting early pregnancy (Zerani et al. 2021). Forming from the remnants of the ovarian follicle after ovulation, its primary function is the production of P4. This hormone is essential during the second half of the menstrual cycle, known as the luteal phase, where the CL becomes the main source of P4 production (Webb et al. 2002; Acosta and Miyamoto 2004; Davis and LaVoie 2019). P4's key role is to prepare the endometrium, the lining of the uterus, for the potential implantation of a fertilized oocyte. By thickening the endometrial lining, it creates a more receptive environment for an embryo. In tandem, the progesterone (P4) helps to maintain the integrity of the endometrial lining (Stouffer 2003; Smith et al. 1975).

In cattle, the initial development of the CL is stimulated by oxytocin, prostaglandins, and P4 until the mid-luteal stage. This proliferation and function of luteal cells is supported by the luteotropic action of various growth factors. The continued high mRNA expression, protein concentration, and cytoplasmic localization of growth factors such as vascular endothelial growth factor (VEGF), fibroblast growth factors (FGF-1 and FGF-2), and insulin-like growth factors (IGFs) in luteal cells during the mid-luteal stage suggest that these factors play a critical role in maintaining luteal cell survival. In the absence of the CL, luteolysis ensues, leading to the termination of pregnancy (Einspanier et al. 1990; Berisha et al. 2002).

The role of EVs in regulating the CL is a growing area of research, shedding light on how these vesicles contribute to luteal function. EVs are known to regulate the lifespan and steroidogenic activity of the CL. They influence the production of P4 and other hormones necessary for maintaining pregnancy (da Silva Rosa et al. 2024a; da Silveira et al. 2012; Capra and Lange-Consiglio 2020; Hart et al. 2022). Notably, EVs from the uterine fluid facilitate communication between various cell types within the CL and other parts of the reproductive system, including the ovary and uterus (Burns et al. 2016; Bridi et al. 2020, 2024). This cross-talk communication can significantly impact the immune environment of the CL, modulating immune cell activity to create a favorable setting for pregnancy (Kusama et al. 2018; Nakamura et al. 2019; Imakawa et al. 2018). By reducing inflammatory responses, these EVs protect luteal cells from immune-mediated damage, establishing a supportive environment for the embryo (Aleksejeva et al. 2022; Nakamura et al. 2020).

Additionally, EVs contribute to angiogenesis, the formation of new blood vessels, which is vital for sustaining hormone production within the CL by ensuring an adequate blood supply (Otsuka et al. 2008; Hamdi et al. 2021). One study on small ffEVs highlighted how P4 levels, modulated by the CL, alter the miRNA content of these EVs. This suggests that the CL not only influences the luteal environment but also has broader effects on ovarian signaling and potentially reproductive outcomes (da Silva Rosa et al. 2024b). Through these molecular signals, EVs regulate key luteal functions, including steroidogenesis and luteal cell survival, reinforcing their essential role in the maintenance of the luteal phase and overall reproductive success.

7 Potential Clinical Applications

The discovery of EVs as key mediators of cell-to-cell communication has opened new avenues for medical research, particularly in the field of reproductive health. The capability to carry a diverse range of bioactive molecules is especially significant in the context of ovarian health, where the loss or dysfunction of specific cell types can severely impair fertility and hormone regulation. Although much remains to be explored regarding the cargo of EVs and the need for standardizing the isolation methods, significant research has been conducted on miRNAs and their potential targets to enhance the understanding of how these vesicles could be used to alleviate or mitigate reproductive disorders. The identification of specific miRNA profiles and their regulatory roles in key signaling pathways offers promising insights into how EVs might be harnessed therapeutically. However, challenges such as optimizing isolation techniques, improving delivery methods, and understanding the biodistribution of EVs in vivo must be addressed to fully unlock their clinical potential in treating ovarian dysfunction and related conditions (Machtinger et al. 2021).

In the medical field, an interesting study by Zhang et al. (2020) demonstrated that EVs derived from human umbilical cord mesenchymal stem cell culture media can be incorporated into cisplatin-damaged GCs, increasing the number of viable cells. This contributes to promoting resistance to apoptosis and restoring the synthesis and secretion of steroid hormones. These effects were observed by noting that E2 levels and the expression of StAR miRNA were higher in the cisplatin-damaged GCs + EVs group compared to the cisplatin-damaged GCs group alone. The authors aimed to test the benefits of using stem cell-derived EVs as a therapeutic strategy for patients with premature ovarian insufficiency (POI), rather than using stem cells directly. Promising results were also reported by Liu et al. (2020), who found that EVs derived from embryonic stem cells helped to recover ovarian function in a chemotherapy-induced premature ovarian failure (POF) model, while also inhibiting GCs apoptosis through activation of the PI3K/AKT pathway. Clonal mesenchymal stem cell-derived EVs were also investigated for their protective effects against GC damage in a study by Eslami et al. (2023). In this well-established chemotherapy-induced ovarian failure model, the authors found a partial recovery of ovarian function. They also suggested that EVs isolated by high-speed centrifugation at 20,000 × g are more cost-effective for potential therapies in patients with POF than conventional EVs isolated at 110,000 × g. However, the authors emphasized the need for further studies to fully understand the cargo and function of these two types of EVs.

ffEVs are also being studied in disease diagnostics. In a proteomics study by Li et al. (2020), ffEVs from polycystic ovaries were found to contain the S100-A9 protein, which promoted inflammation by activating the NF-κB pathway. Similarly, Wang et al. (2021b) identified abnormal expression of long-coding RNA in the ffEVs of patients with polycystic ovary syndrome (PCOS). In this context, miRNAs are emerging as potential biomarkers for the diagnosis and treatment of PCOS. Yang et al. (2023) discovered that various miRNAs and proteins are differently expressed in ffEVs from women with PCOS, including GRAMD1B and SPTLC2 (proteins) and miR-206 and miR-204-5p (miRNAs), which are involved in hormone metabolism pathways.

Alternatively, the transplantation of EVs from human-induced pluripotent mesenchymal stem cells into mouse GCs resulted in significant restoration of ovarian follicle numbers and enhanced cell proliferation in a study conducted by Cao et al. (2023). This treatment led to upregulation of the integrin-linked kinase (ILK)-PI3K/AKT pathway, facilitated by the transfer of regulatory miRNAs that target ILK pathway genes. In contrast, mural GCs from women with PCOS exhibited greater differences in miRNA content compared to EVs from both healthy and PCOS groups, where these differences were less pronounced (Rooda et al. 2020). The authors suggested that these miRNA expression alterations could disrupt E2 receptor signaling and contribute to the dysregulation of transcription and apoptosis. Additionally, exosomal miR-424-5p derived from PCOS FF has been shown to inhibit

GC proliferation and induce senescence by targeting CDCA4-mediated Rb/E2F1 signaling (Yuan et al. 2021). Some studies have also been conducted to investigate the effects of EVs on CCs. In a study performed by Zhou et al. (2022), CCs collected from mice via ovariohysterectomy experienced growth inhibition when co-cultured with adipocytes, which stimulated apoptosis through a paracrine effect. The key findings revealed that miR-26b was upregulated in EVs derived from mature adipocytes, and these EVs inhibited CC viability by targeting the *Jag1* gene. Furthermore, these EVs disrupted ovulation and altered hormone levels in vivo, contributing to the development of PCOS.

Although more studies are necessary, ovarian-derived EVs hold significant potential for clinical applications, particularly in addressing conditions like POI, PCOS, and ovarian failure. By mediating critical processes such as CG recovery, hormone regulation, and apoptosis inhibition, these vesicles offer a novel therapeutic approach that bypasses the complexities of direct stem cell therapies. Further research into the cargo and functions of these EVs will be crucial in developing effective and cost-efficient treatments for ovarian dysfunction, providing hope for patients with reproductive challenges.

8 Conclusions

Extracellular vesicles are emerging as pivotal players in ovarian physiology, facilitating essential communication between oocytes, granulosa cells, cumulus cells, and the ovarian microenvironment. They play a key role in regulating processes such as oocyte maturation, folliculogenesis, and stress response, helping to ensure optimal oocyte competence and overall reproductive success. By carrying bioactive molecules like miRNAs, proteins, and lipids, EVs can influence gene expression in both granulosa cells and cumulus cells, supporting cell proliferation, stress resistance, and hormone regulation. These vesicles also modulate interactions with immune cells and hormonal environments, further enhancing their impact on ovarian health. Their emerging role in diagnostics also holds potential for developing biomarkers to detect ovarian pathologies early; however, continued research into the mechanisms of EV-mediated communication and their therapeutic potential is essential to further unlock their utility in enhancing fertility treatments and addressing reproductive disorders.

Acknowledgments

Authors would like to thank the São Paulo Research Foundation (FAPESP) for the scholarships awarded to Matheus A. Chaves (22/01236-7), Juliana G. Ferst (20/13075-2 and 23/15072-9), Mariani F. Fiorenza (22/01235-0) and Franciele F. Vit (22/02701-5). Juliano C. da Silveira also thanks FAPESP for the financial support (21/06645-0) and the National Council for Scientific and Technological Development (CNPq, grant number #308101/2021-9).

Compliance with Ethical Standards Authors declare that they have no conflict of interest. This chapter does not contain any studies with animals performed by any of the authors.

References

Abdulrahman N, Fair T (2019) Contribution of the immune system to follicle differentiation, ovulation and early corpus luteum formation. Anim Reprod 16:440–448. https://doi.org/10.21451/1984-3143-AR2019-0087

Abumaghaid MM, Abdelazim AM, Belali TM et al (2022) Shuttle transfer of mRNA transcripts via extracellular vesicles from male reproductive tract cells to the cumulus–oocyte complex in rabbits (Oryctolagus cuniculus). Front Vet Sci 9:2022. https://doi.org/10.3389/fvets.2022.816080

Acosta TJ, Miyamoto A (2004) Vascular control of ovarian function: ovulation, corpus luteum formation and regression. Anim Reprod Sci 82–83:127–140. https://doi.org/10.1016/j.anireprosci.2004.04.022

Adams GP, Jaiswal R, Singh J, Malhi P (2008) Progress in understanding ovarian follicular dynamics in cattle. Theriogenology 69:72–80. https://doi.org/10.1016/j.theriogenology.2007.09.026

Ahmed TA, Ahmed SM, El-Gammal Z et al (2020) Oocyte aging: the role of cellular and environmental factors and impact on female fertility. In: Advances in experimental medicine and biology. Springer, pp 109–123. https://doi.org/10.1007/5584_2019_456

Aleksejeva E, Zarovni N, Dissanayake K et al (2022) Extracellular vesicle research in reproductive science: paving the way for clinical achievements. Biol Reprod 106:408–424. https://doi.org/10.1093/biolre/ioab245

Andrade GM, Meirelles FV, Perecin F, da Silveira JC (2017) Cellular and extracellular vesicular origins of miRNAs within the bovine ovarian follicle. Reprod Domest Anim 52:1036–1045. https://doi.org/10.1111/rda.13021

Aparicio IM, Garcia-Herreros M, O'Shea LC et al (2011) Expression, regulation, and function of progesterone receptors in bovine cumulus oocyte complexes during in vitro maturation. Biol Reprod 84:910–921. https://doi.org/10.1095/biolreprod.110.087411

Arici A, Oral E, Bahtiyar O et al (1997) Leukaemia inhibitory factor expression in human follicular fluid and ovarian cells. Hum Reprod 12:1233–1239. https://doi.org/10.1093/humrep/12.6.1233

Asaadi A, Dolatabad NA, Atashi H et al (2021) Extracellular vesicles from follicular and ampullary fluid isolated by density gradient ultracentrifugation improve bovine embryo development and quality. Int J Mol Sci 22:1–16. https://doi.org/10.3390/ijms22020578

Battaglia R, Musumeci P, Ragusa M et al (2020) Ovarian aging increases small extracellular vesicle CD81+ release in human follicular fluid and influences miRNA profiles. Aging (Albany NY) 12:12324–12341. https://doi.org/10.18632/aging.103441

Benedetti C, Pavani KC, Gansemans Y et al (2024) From follicle to blastocyst: microRNA-34c from follicular fluid-derived extracellular vesicles modulates blastocyst quality. J Anim Sci Biotechnol 15:104. https://doi.org/10.1186/s40104-024-01059-8

Berisha B, Schams D, Miyamoto A (2002) The expression of angiotensin and endothelin system members in bovine corpus luteum during estrous cycle and pregnancy. Endocrine 19:305–312. https://doi.org/10.1385/ENDO:19:3:305

Bisinotto RS, Chebel RC, Santos JEP (2010) Follicular wave of the ovulatory follicle and not cyclic status influences fertility of dairy cows. J Dairy Sci 93:3578–3587. https://doi.org/10.3168/jds.2010-3047

Branham KL (2015) Role of luteal cell-derived exosomes in communication with immune cells. Master thesis, PennState University

Bridi A, Perecin F, da Silveira JC (2020) Extracellular vesicles mediated early embryo–maternal interactions. Int J Mol Sci 21. https://doi.org/10.3390/ijms21031163
Bridi A, Sangalli JR, Nociti RP et al (2024) Small extracellular vesicles derived from the crosstalk between early embryos and the endometrium potentially mediate corpus luteum function. Biol Reprod:ioae143. https://doi.org/10.1093/biolre/ioae143
Brinca AT, Ramalhinho AC, Sousa Â et al (2022) Follicular fluid: a powerful tool for the understanding and diagnosis of polycystic ovary syndrome. Biomedicine 10:1254. https://doi.org/10.3390/biomedicines10061254
Brinca AT, Peiró AM, Evangelio PM et al (2024) Follicular fluid and blood monitorization of infertility biomarkers in women with endometriosis. Int J Mol Sci 25:7177. https://doi.org/10.3390/ijms25137177
Bukovský A, Presl J, Krabec Z, Bednarík T (1977) Ovarian function in adult rats treated with antithymocyte serum. Experientia 33:280–281. https://doi.org/10.1007/BF02124112
Bulmer D (1964) The histochemistry of ovarian macrophages in the rat. J Anat 98:313–319
Burns GW, Brooks KE, Spencer TE (2016) Extracellular vesicles originate from the conceptus and uterus during early pregnancy in sheep. Biol Reprod 94:1–11. https://doi.org/10.1095/biolreprod.115.134973
Cao RC, Lv Y, Lu G et al (2023) Extracellular vesicles from iPSC-MSCs alleviate chemotherapy-induced mouse ovarian damage via the ILK-PI3K/AKT pathway. Zool Res 44:620. https://doi.org/10.24272/j.issn.2095-8137.2022.340
Capra E, Lange-Consiglio A (2020) The biological function of extracellular vesicles during fertilization, early embryo – maternal crosstalk and their involvement in reproduction: review and overview. Biomol Ther 10:1–26. https://doi.org/10.3390/biom10111510
Capra E, Kosior MA, Cocchia N et al (2022) Variations of follicular fluid extracellular vesicles miRNAs content in relation to development stage and season in buffalo. Sci Rep 12:14886. https://doi.org/10.1038/s41598-022-18438-8
Care AS, Diener KR, Jasper MJ et al (2013) Macrophages regulate corpus luteum development during embryo implantation in mice. J Clin Invest 123:3472–3487. https://doi.org/10.1172/JCI60561
Chen L, Russell PT, Larsen WJ (1993) Functional significance of cumulus expansion in the mouse: roles for the preovulatory synthesis of hyaluronic acid within the cumulus mass. Mol Reprod Dev 34:87–93. https://doi.org/10.1002/mrd.1080340114
Collodel G, Gambera L, Stendardi A et al (2023) Follicular fluid components in reduced ovarian reserve, endometriosis, and idiopathic infertility. Int J Mol Sci 24:2589. https://doi.org/10.3390/ijms24032589
da Silva Rosa PM, Bridi A, de Ávila FG et al (2024a) Corpus luteum proximity alters molecular signature of the small extracellular vesicles and cumulus cells in the bovine ovarian follicle environment. Mol Cell Endocrinol 592. https://doi.org/10.1016/j.mce.2024.112347
da Silva Rosa PM, Bridi A, de Ávila FG et al (2024b) Corpus luteum presence in the bovine ovary increase intrafollicular progesterone concentration: consequences in follicular cells gene expression and follicular fluid small extracellular vesicles miRNA contents. J Ovarian Res 17:1–14. https://doi.org/10.1186/s13048-024-01387-3
da Silveira JC, Veeramachaneni DNR, Winger QA et al (2012) Cell-secreted vesicles in equine ovarian follicular fluid contain mirnas and proteins: a possible new form of cell communication within the ovarian follicle. Biol Reprod 86:1–10. https://doi.org/10.1095/biolreprod.111.093252
Da Silveira JC, De Andrade GM, Nogueira MFG et al (2015) Involvement of miRNAs and cell-secreted vesicles in mammalian ovarian antral follicle development. Reprod Sci 22:1474–1483. https://doi.org/10.1177/1933719115574344
Da Silveira JC, Andrade GM, Collado M d et al (2017) Supplementation with small-extracellular vesicles from ovarian follicular fluid during in vitro production modulates bovine embryo development. PLoS One 12:e0179451. https://doi.org/10.1371/journal.pone.0179451

Da Silveira JC, Andrade GM, Simas RC et al (2021) Lipid profile of extracellular vesicles and their relationship with bovine oocyte developmental competence: new players in intra follicular cell communication. Theriogenology 174:1–8. https://doi.org/10.1016/j.theriogenology.2021.07.024
Dalanezi FM, Garcia HDM, Ferrazza FRA et al (2019) Extracellular vesicles of follicular fluid from heat-stressed cows modify the gene expression of in vitro-matured oocytes. Anim Reprod Sci 205:94–104. https://doi.org/10.1016/j.anireprosci.2019.04.008
Davis JS, LaVoie HA (2019) Molecular regulation of progesterone production in the corpus luteum. In: The ovary, 3rd edn. Elsevier Inc, pp 237–253
De Ávila ACFCM, Da Silveira JC (2020) Role of extracellular vesicles during oocyte maturation and early embryo development. Reprod Fertil Dev 32:56–64. https://doi.org/10.1071/RD19389
De Ávila ACFCM, Bridi A, Andrade GM et al (2020) Estrous cycle impacts microRNA content in extracellular vesicles that modulate bovine cumulus cell transcripts during in vitro maturation. Biol Reprod 102:362–375. https://doi.org/10.1093/biolre/ioz177
Denicol AC, Lopes J, Mendonça LGD et al (2012) Low progesterone concentration during the development of the first follicular wave reduces pregnancy per insemination of lactating dairy cows. J Dairy Sci 95:1794–1806. https://doi.org/10.3168/jds.2011-4650
Diez-Fraile A, Lammens T, Tilleman K et al (2014) Age-associated differential microRNA levels in human follicular fluid reveal pathways potentially determining fertility and success of in vitro fertilization. Hum Fertil 17:90–98. https://doi.org/10.3109/14647273.2014.897006
Duncan A, Forcina J, Birt A, Townson D (2012) Estrous cycle-dependent changes of Fas expression in the bovine corpus luteum: influence of keratin 8/18 intermediate filaments and cytokines. Reprod Biol Endocrinol 10:1. https://doi.org/10.1186/1477-7827-10-90
Edure T, Matsuno Y, Matsushita K et al (2024) Dynamics of extracellular vesicle uptake by mural granulosa cells in mice. Mol Reprod Dev 91:e23737. https://doi.org/10.1002/mrd.23737
Einspanier R, Miyamoto A, Schams D et al (1990) Tissue concentration , mRNA expression and stimulation of IGF-I in luteal tissue during the oestrous cycle and pregnancy of cows. J Reprod Fertil 90:439–445. https://doi.org/10.1530/jrf.0.0900439
Enien WM, Chantler E, Seif MW, Elstein M (1998) Human ovarian granulosa cells and follicular fluid indices: the relationship to oocyte maturity and fertilization in vitro. Hum Reprod 13:1303–1306. https://doi.org/10.1093/humrep/13.5.1303
Eslami N, Bahrehbar K, Esfandiari F et al (2023) Regenerative potential of different extracellular vesicle subpopulations derived from clonal mesenchymal stem cells in a mouse model of chemotherapy-induced premature ovarian failure. Life Sci 321:121536. https://doi.org/10.1016/j.lfs.2023.121536
Fainaru O, Amsalem H, Bentov Y et al (2010) CD56brightCD16-natural killer cells accumulate in the ovarian follicular fluid of patients undergoing in vitro fertilization. Fertil Steril 94:1918–1921. https://doi.org/10.1016/j.fertnstert.2009.12.067
Fiorentino G, Merico V, Zanoni M et al (2024) Extracellular vesicles secreted by cumulus cells contain microRNAs that are potential regulatory factors of mouse oocyte developmental competence. Mol Hum Reprod 30:gaae019. https://doi.org/10.1093/molehr/gaae019
Gabryś J, Kij-Mitka B, Sawicki S et al (2022) Extracellular vesicles from follicular fluid may improve the nuclear maturation rate of in vitro matured mare oocytes. Theriogenology 188:116–124. https://doi.org/10.1016/j.theriogenology.2022.05.022
Gabryś J, Gurgul A, Szmatoła T et al (2024) Follicular fluid-derived extracellular vesicles influence on in vitro maturation of equine oocyte: impact on cumulus cell viability, expansion and transcriptome. Int J Mol Sci 25:3262. https://doi.org/10.3390/ijms25063262
Gad A, Murin M, Bartkova A et al (2022) Small-extracellular vesicles and their microRNA cargo from porcine follicular fluids: the potential association with oocyte quality. J Anim Sci Biotechnol 13:82. https://doi.org/10.1186/s40104-022-00723-1
Gebremedhn S, Gad A, Aglan HS et al (2020) Extracellular vesicles shuttle protective messages against heat stress in bovine granulosa cells. Sci Rep 10:15824. https://doi.org/10.1038/s41598-020-72706-z

Grzesiak M, Popiolek K, Knapczyk-Stwora K (2020) Extracellular vesicles in follicular fluid of sexually mature gilts' ovarian antral follicles - identification and proteomic analysis. J Physiol Pharmacol 71:10.26402/jpp.2020.1.13

Gurunathan S, Kang MH, Song H et al (2022) The role of extracellular vesicles in animal reproduction and diseases. J Anim Sci Biotechnol 13. https://doi.org/10.1186/s40104-022-00715-1

Hailay T, Hoelker M, Poirier M et al (2019) Extracellular vesicle-coupled miRNA profiles in follicular fluid of cows with divergent post-calving metabolic status. Sci Rep 9:12851. https://doi.org/10.1038/s41598-019-49029-9

Hamdi M, Cañon-Beltrán K, Mazzarella R et al (2021) Characterization and profiling analysis of bovine oviduct and uterine extracellular vesicles and their miRNA cargo through the estrous cycle. FASEB J 35:1–15. https://doi.org/10.1096/fj.202101023R

Harris EA, Stephens KK, Winuthayanon W (2020) Extracellular vesicles and the oviduct function. Int J Mol Sci 21:1–20. https://doi.org/10.3390/ijms21218280

Hart AR, Khan NLA, Godakumara K et al (2022) The role of extracellular vesicles in endometrial receptivity and their potential in reproductive therapeutics and diagnosis. Reprod Biol 22: 100645. https://doi.org/10.1016/j.repbio.2022.100645

Hasan MM, Viil J, Lättekivi F et al (2020) Bovine follicular fluid and extracellular vesicles derived from follicular fluid alter the bovine oviductal epithelial cells transcriptome. Int J Mol Sci 21:1–16. https://doi.org/10.3390/ijms21155365

Hehnke KE, Christenson LK, Ford SP, Taylor M (1994) Macrophage infiltration into the porcine corpus luteum during prostaglandin F(2α)-induced luteolysis. Biol Reprod 50:10–15. https://doi.org/10.1095/biolreprod50.1.10

Hung WT, Hong X, Christenson LK, McGinnis LK (2015) Extracellular vesicles from bovine follicular fluid support cumulus expansion. Biol Reprod 93:117. https://doi.org/10.1095/biolreprod.115.132977

Hung WT, Navakanitworakul R, Khan T et al (2017) Stage-specific follicular extracellular vesicle uptake and regulation of bovine granulosa cell proliferation. Biol Reprod 97:644–655. https://doi.org/10.1093/biolre/iox106

Imakawa K, Bai R, Kusama K (2018) Integration of molecules to construct the processes of conceptus implantation to the maternal endometrium. J Anim Sci 96:3009–3021. https://doi.org/10.1093/jas/sky103

Javadi M, Soleimani Rad J, Pashaiasl M et al (2022) The effects of plasma-derived extracellular vesicles on cumulus expansion and oocyte maturation in mice. Reprod Biol 22:100593. https://doi.org/10.1016/j.repbio.2021.100593

Jiemtaweeboon S, Shirasuna K, Nitta A et al (2011) Evidence that polymorphonuclear neutrophils infiltrate into the developing corpus luteum and promote angiogenesis with interleukin-8 in the cow. Reprod Biol Endocrinol 9:1–10. https://doi.org/10.1186/1477-7827-9-79

Korzekwa A, Murakami S, Wocławek-Potocka I et al (2008) The influence of tumor necrosis factor α (TNF) on the secretory function of bovine corpus luteum: TNF and its receptors expression during the estrous cycle. Reprod Biol 8:245–262. https://doi.org/10.1016/s1642-431x(12)60015-1

Kusama K, Nakamura K, Bai R et al (2018) Intrauterine exosomes are required for bovine conceptus implantation. Biochem Biophys Res Commun 495:1370–1375. https://doi.org/10.1016/j.bbrc.2017.11.176

Lawler DF, Hopkins J, Watson ED (1999) Immune cell populations in the equine corpus luteum throughout the oestrous cycle and early pregnancy: an immunohistochemical and flow cytometric study. J Reprod Fertil 117:281–290. https://doi.org/10.1530/jrf.0.1170281

Lee SH, Oh HJ, Kim MJ, Lee BC (2020) Exosomes derived from oviduct cells mediate the EGFR/MAPK signaling pathway in cumulus cells. J Cell Physiol 235:1386–1404. https://doi.org/10.1002/jcp.29058

Lei ZM, Chegini N, Rao CV (1991) Quantitative cell composition of human and bovine corpora lutea from various reproductive states. Biol Reprod 1156:1148–1156. https://doi.org/10.1095/biolreprod44.6.1148
Li H, Huang X, Chang X et al (2020) S100-A9 protein in exosomes derived from follicular fluid promotes inflammation via activation of NF-κB pathway in polycystic ovary syndrome. J Cell Mol Med 24:114–125. https://doi.org/10.1111/jcmm.14642
Li X, Du M, Liu Y et al (2024) Proteome and metabolomic profile of Mongolian horse follicular fluid during follicle development. Sci Rep 14:19788. https://doi.org/10.1038/s41598-024-66686-7
Liu M, Qiu Y, Xue Z et al (2020) Small extracellular vesicles derived from embryonic stem cells restore ovarian function of premature ovarian failure through PI3K/AKT signaling pathway. Stem Cell Res Ther 11:1–12. https://doi.org/10.1186/s13287-019-1508-2
Macaulay AD, Gilbert I, Caballero J et al (2014) The gametic synapse: RNA transfer to the bovine oocyte. Biol Reprod 91:1–12. https://doi.org/10.1095/biolreprod.114.119867
Martinez RM, Liang L, Racowsky C et al (2018) Extracellular microRNAs profile in human follicular fluid and IVF outcomes. Sci Rep 8:17036. https://doi.org/10.1038/s41598-018-35379-3
Martinez RM, Baccarelli AA, Liang L et al (2019) Body mass index in relation to extracellular vesicle–linked microRNAs in human follicular fluid. Fertil Steril 112:387–396.e3. https://doi.org/10.1016/j.fertnstert.2019.04.001
Matoba S, Bender K, Fahey AG et al (2014) Predictive value of bovine follicular components as markers of oocyte developmental potential. Reprod Fertil Dev 26:337–345. https://doi.org/10.1071/RD13007
Menjivar NG, Gad A, Gebremedhn S et al (2023) Granulosa cell-derived extracellular vesicles mitigate the detrimental impact of thermal stress on bovine oocytes and embryos. Front Cell Dev Biol 11:2023. https://doi.org/10.3389/fcell.2023.1142629
Montani DA, Braga DP d AF, Borges E et al (2019) Understanding mechanisms of oocyte development by follicular fluid lipidomics. J Assist Reprod Genet 36:1003–1011. https://doi.org/10.1007/s10815-019-01428-7
Machtinger R, Baccarelli AA, Wu H (2021) Extracellular vesicles and female reproduction. J Assist Reprod Genet 38:549–557. https://doi.org/10.1007/s10815-020-02048-2
Moreira MV, Vale-Fernandes E, Albergaria IC et al (2023) Follicular fluid composition and reproductive outcomes of women with polycystic ovary syndrome undergoing in vitro fertilization: a systematic review. Rev Endocr Metab Disord 24:1045–1073. https://doi.org/10.1007/s11154-023-09819-z
Nakamura K, Kusama K, Ideta A et al (2019) Effects of miR-98 in intrauterine extracellular vesicles on maternal immune regulation during the peri-implantation period in cattle. Sci Rep 9:1–10. https://doi.org/10.1038/s41598-019-56879-w
Nakamura K, Kusama K, Suda Y et al (2020) Emerging role of extracellular vesicles in embryo–maternal communication throughout implantation processes. Int J Mol Sci 21:1–15. https://doi.org/10.3390/ijms21155523
Navakanitworakul R, Hung WT, Gunewardena S et al (2016) Characterization and small RNA content of extracellular vesicles in follicular fluid of developing bovine antral follicles. Sci Rep 6:25486. https://doi.org/10.1038/srep25486
Nejabati HR, Roshangar L, Nouri M (2023) Follicular fluid extracellular vesicle miRNAs and ovarian aging. Clin Chim Acta 538:29–35. https://doi.org/10.1016/j.cca.2022.11.003
Neuvians TP, Schams D, Berisha B, Pfaffl MW (2004) Involvement of pro-inflammatory cytokines, mediators of inflammation, and basic fibroblast growth factor in prostaglandin F2α-induced luteolysis in bovine corpus luteum. Biol Reprod 70:473–480. https://doi.org/10.1095/biolreprod.103.016154
O'Brien Y, Wingfield M, O'Shea LC (2019) Anti-Müllerian hormone and progesterone levels in human follicular fluid are predictors of embryonic development. Reprod Biol Endocrinol 17:47. https://doi.org/10.1186/s12958-019-0492-9

O'Shea LC, Daly E, Hensey C, Fair T (2017) ATRX is a novel progesterone-regulated protein and biomarker of low developmental potential in mammalian oocytes. Reproduction 153:671–682. https://doi.org/10.1530/REP-16-0443
Oakley OR, Kim HY, El-Amouri I et al (2010) Periovulatory leukocyte infiltration in the rat ovary. Endocrinology 151:4551–4559. https://doi.org/10.1210/en.2009-1444
Okuda K, Sakumoto R, Uenoyama Y et al (1999) Tumor necrosis factor α receptors in microvascular endothelial cells from bovine corpus luteum. Biol Reprod 61:1017–1022. https://doi.org/10.1095/biolreprod61.4.1017
Otsuka M, Zheng M, Hayashi M et al (2008) Impaired microRNA processing causes corpus luteum insufficiency and infertility in mice. J Clin Invest 118:1944–1954. https://doi.org/10.1172/JCI33680
Paavola LG (1979) The corpus luteum of the Guinea pig. IV. Fine structure of macrophages during pregnancy and postpartum luteolysis, and the phagocytosis of luteal cells. Am J Anat 154:337–363. https://doi.org/10.1002/aja.1001540304
Peluso JJ, Liu X, Gawkowska A et al (2010) Progesterone inhibits apoptosis in part by PGRMC1-regulated gene expression. Mol Cell Endocrinol 320:153–161. https://doi.org/10.1016/j.mce.2010.02.005
Penny LA, Armstrong D, Bramley TA et al (1999) Immune cells and cytokine production in the bovine corpus luteum throughout the oestrous cycle and after induced luteolysis. J Reprod Fertil 115:87–96. https://doi.org/10.1530/jrf.0.1150087
Pocate-Cheriet K, Santulli P, Kateb F et al (2020) The follicular fluid metabolome differs according to the endometriosis phenotype. Reprod Biomed Online 41:1023–1037. https://doi.org/10.1016/j.rbmo.2020.09.002
Poulsen L l C, Pla I, Sanchez A et al (2019) Progressive changes in human follicular fluid composition over the course of ovulation: quantitative proteomic analyses. Mol Cell Endocrinol 495:110522. https://doi.org/10.1016/j.mce.2019.110522
Reader KL, Stanton JAL, Juengel JL (2017) The role of oocyte organelles in determining developmental competence. Biology (Basel) 6. https://doi.org/10.3390/biology6030035
Reibiger I, Spanel-Borowski K (2000) Difference in localization of eosinophils and mast cells in the bovine ovary. J Reprod Fertil 118:243–249
Reynolds LP, Redmer DA (1999) Growth and development of the corpus luteum. J Reprod Fertil Suppl 54:181–191
Rodosthenous RS, Baccarelli AA, Mansour A et al (2019) Supraphysiological concentrations of bisphenol a alter the expression of extracellular vesicle-enriched miRNAs from human primary granulosa cells. Toxicol Sci 169:5–13. https://doi.org/10.1093/toxsci/kfz020
Rodrigues TA, Tuna KM, Alli AA et al (2019) Follicular fluid exosomes act on the bovine oocyte to improve oocyte competence to support development and survival to heat shock. Reprod Fertil Dev 31:888–897. https://doi.org/10.1071/RD18450
Rooda I, Hasan MM, Roos K et al (2020) Cellular, extracellular and extracellular vesicular miRNA profiles of pre-ovulatory follicles indicate signaling disturbances in polycystic ovaries. Int J Mol Sci 21:9550. https://doi.org/10.3390/ijms21249550
Saeed-Zidane M, Linden L, Salilew-Wondim D et al (2017) Cellular and exosome mediated molecular defense mechanism in bovine granulosa cells exposed to oxidative stress. PLoS One 12:e0187569. https://doi.org/10.1371/journal.pone.0187569
Sirard MA, Richard F, Blondin P, Robert C (2006) Contribution of the oocyte to embryo quality. Theriogenology 65:126–136. https://doi.org/10.1016/j.theriogenology.2005.09.020
Smith SM, Freeman ME, Neill JD (1975) The control of progesterone secretion during the estrous cycle and early pseudopregnancy in the rat: prolactin, gonadotropin and steroid levels associated with Rescue of the Corpus Luteum of Pseudopregnancy12. Endocrinology 96:219–226. https://doi.org/10.1210/endo-96-1-219
Smith MP, Flannery GR, Randle BJ et al (2005) Leukocyte origin and profile in follicular aspirates at oocyte retrieval. Hum Reprod 20:3526–3531. https://doi.org/10.1093/humrep/dei240

Sohel MMH, Hoelker M, Noferesti SS et al (2013) Exosomal and non-exosomal transport of extracellular microRNAs in follicular fluid: implications for bovine oocyte developmental competence. PLoS One 8:e78505. https://doi.org/10.1371/journal.pone.0078505
Spanel-Borowski K, Rahner P, Ricken AM (1997) Immunolocalization of CD18-positive cells in the bovine ovary. J Reprod Fertil 111:197–205. https://doi.org/10.1530/jrf.0.1110197
Stouffer RL (2003) Progesterone as a mediator of gonadotrophin action in the corpus luteum: beyond steroidogenesis. Hum Reprod Update 9:99–117. https://doi.org/10.1093/humupd/dmg016
Sysoeva AP, Makarova NP, Silachev DN et al (2021) Influence of extracellular vesicles of the follicular fluid on morphofunctional characteristics of human sperm. Bull Exp Biol Med 172: 254–262. https://doi.org/10.1007/s10517-021-05372-4
Tesfaye D, Hailay T, Salilew-Wondim D et al (2020) Extracellular vesicle mediated molecular signaling in ovarian follicle: implication for oocyte developmental competence. Theriogenology 150:70–74. https://doi.org/10.1016/j.theriogenology.2020.01.075
Townson DH, Liptak AR (2003) Chemokines in the corpus luteum: implications of leukocyte chemotaxis. Reprod Biol Endocrinol 1:1–7. https://doi.org/10.1186/1477-7827-1-94
Townson DH, O'Connor CL, Pru JK (2002) Expression of monocyte chemoattractant protein-1 and distribution of immune cell populations in the bovine corpus luteum throughout the estrous cycle. Biol Reprod 66:361–366. https://doi.org/10.1095/biolreprod66.2.361
Uzbekova S, Almiñana C, Labas V et al (2020) Protein cargo of extracellular vesicles from bovine follicular fluid and analysis of their origin from different ovarian cells. Front Vet Sci 7:584948. https://doi.org/10.3389/fvets.2020.584948
Wada SI, Kudo T, Kudo M et al (1999) Induction of macrophage migration inhibitory factor in human ovary by human chorionic gonadotrophin. Hum Reprod 14:395–399. https://doi.org/10.1093/humrep/14.2.395
Walter J, Monthoux C, Fortes C et al (2020) The bovine cumulus proteome is influenced by maturation condition and maturational competence of the oocyte. Sci Rep 10:9880. https://doi.org/10.1038/s41598-020-66822-z
Wang X, Meng K, Wang H et al (2021a) Identification of small extracellular vesicle subtypes in follicular fluid: insights into the function and miRNA profiles. J Cell Physiol 236:5633–5645. https://doi.org/10.1002/jcp.30251
Wang L, Fan H, Zou Y et al (2021b) Aberrant expression of long non-coding RNAs in exosomes in follicle fluid from PCOS patients. Front Genet 11:608178. https://doi.org/10.3389/fgene.2020.608178
Wang X, Zhang Z, Qi Y et al (2024) Study of the uptake mechanism of two small extracellular vesicle subtypes by granulosa cells. Anim Reprod Sci 270:107576. https://doi.org/10.1016/j.anireprosci.2024.107576
Webb R, Woad KJ, Armstrong DG (2002) Corpus luteum (CL) function: Local control mechanisms. Domest Anim Endocrinol 23:277–285. https://doi.org/10.1016/S0739-7240(02)00163-7
Wu R, Van der Hoek KH, Ryan NK et al (2004) Macrophage contributions to ovarian function. Hum Reprod Update 10:119–133. https://doi.org/10.1093/humupd/dmh011
Xiao Y, Peng X, Peng Y et al (2022) Macrophage-derived extracellular vesicles regulate follicular activation and improve ovarian function in old mice by modulating local environment. Clin Transl Med 12:e1071. https://doi.org/10.1002/ctm2.1071
Yang X, Gilman-Sachs A, Kwak-Kim J (2019) Ovarian and endometrial immunity during the ovarian cycle. J Reprod Immunol 133:7–14. https://doi.org/10.1016/j.jri.2019.04.001
Yang Y, Lang P, Zhang X et al (2023) Molecular characterization of extracellular vesicles derived from follicular fluid of women with and without PCOS: integrating analysis of differential miRNAs and proteins reveals vital molecules involving in PCOS. J Assist Reprod Genet 40:537–552. https://doi.org/10.1007/s10815-023-02724-z
Ying W, Yunqi Z, Zimeng L et al (2023) Large extracellular vesicles in bovine follicular fluid inhibit the apoptosis of granulosa cell and stimulate the production of steroid hormones. Theriogenology 195:149–158. https://doi.org/10.1016/j.theriogenology.2022.10.027

Ying W, Yunqi Z, Deji L et al (2024) Follicular fluid HD-sevs-mir-128–3p is a key molecule in regulating bovine granulosa cells autophagy. Theriogenology 226:263–276. https://doi.org/10.1016/j.theriogenology.2024.06.022

Yuan D, Luo J, Sun Y et al (2021) PCOS follicular fluid derived exosomal miR-424-5p induces granulosa cells senescence by targeting CDCA4 expression. Cell Signal 85:110030. https://doi.org/10.1016/j.cellsig.2021.110030

Zerani M, Polisca A, Boiti C et al (2021) Current knowledge on the multifactorial regulation of corpora lutea lifespan: the rabbit model. Animals 11:1–18. https://doi.org/10.3390/ani11020296

Zhang D, Lv J, Tang R et al (2021) Association of exosomal microRNAs in human ovarian follicular fluid with oocyte quality. Biochem Biophys Res Commun 534:468–473. https://doi.org/10.1016/j.bbrc.2020.11.058

Zhang J, Yin H, Jiang H et al (2020) The protective effects of human umbilical cord mesenchymal stem cell-derived extracellular vesicles on cisplatin-damaged granulosa cells. Taiwan J Obstet Gynecol 59:527–533. https://doi.org/10.1016/j.tjog.2020.05.010

Zheng Y, Chen KL, Zheng XM et al (2014) Identification and bioinformatics analysis of microRNAs associated with stress and immune response in serum of heat-stressed and normal Holstein cows. Cell Stress Chaperones 19:973–981. https://doi.org/10.1007/s12192-014-0521-8

Zhou CJ, Wu SN, Shen JP et al (2016) The beneficial effects of cumulus cells and oocyte-cumulus cell gap junctions depends on oocyte maturation and fertilization methods in mice. PeerJ 4:e1761. https://doi.org/10.7717/peerj.1761

Zhou G, Gu Y, Zhou F et al (2022) Adipocytes-derived extracellular vesicle-miR-26b promotes apoptosis of cumulus cells and induces polycystic ovary syndrome. Front Endocrinol 12:789939. https://doi.org/10.3389/fendo.2021.789939

Extracellular Vesicles in Implantation: Cross-Talk Between the Embryo and Endometrium

Muhammad Ashir Nabeel and Romana A. Nowak

Contents

Abstract Extracellular vesicles (EVs), including exosomes and microvesicles, have emerged as pivotal mediators of intercellular communication. Embryo implantation is a critical process in early pregnancy and requires communication between the embryo and maternal uterus. EVs are important in coordinating the communication between the embryo and maternal uterus. This review explores EV biogenesis,

M. A. Nabeel and R. A. Nowak (✉)
Department of Animal Sciences, College of Agricultural, Consumer and Environmental Sciences, University of Illinois, Urbana, IL, USA
e-mail: mnabeel2@illinois.edu; ranowak@illinois.edu

Adv Anat Embryol Cell Biol (2025) 241: 147–170

https://doi.org/10.1007/102_2024_8,

Published Online: 01 January 2025

molecular composition, and functional roles during implantation. It emphasizes the dynamic role of EVs in modulating the maternal-embryo dialogue, which is critical for establishing a receptive endometrium and facilitating successful implantation. EVs secreted by the embryo and endometrial cells have been shown to carry a diverse cargo of proteins, lipids, and miRNAs, which collectively influence key physiological processes, including immune tolerance, endometrial receptivity, and trophoblast invasion. EVs can be potential candidates as non-invasive biomarkers to assess the quality of embryos and uterine receptivity to enhance reproductive success. By providing a comprehensive overview of the current understanding of EVs in implantation, this chapter aims to highlight the significance of EVs in reproductive biology and their potential applications in improving fertility rates.

Keywords Conceptus, Endometrium, Extracellular vesicles, Implantation, Internalization, miRNAs, Proteins, Trophectoderm, Uterine fluid, Uterine receptivity

Abbreviations

AKR1B1	Aldo-keto reductase family 1 member B1
AI	Artificial insemination
bEECs	Bovine endometrial epithelial cells
JNK	c-Jun N-terminal kinase
dESCs	Decidualized endometrial stromal cells
EVs	Extracellular vesicles
EECs	Endometrial epithelial cells
ES	Embryonic stem cells
EEC-1	Endometrial adenocarcinoma cell
sFRP2	Frizzled-related protein 2
FAK	Focal adhesion kinase
HEE	Human endometrial epithelial cells
HUVECs	Human umbilical vein endothelial cells
endMSCs	Endometrial mesenchymal stromal cells
IVF	In vitro fertilized
IVP	In vitro produced
IVV	In vivo produced
ICM/TE	Inner cell mass/trophectoderm
IVC	In vitro culture
ISGs	Interferon-stimulated genes
IFNT	Interferon-tau
miRNAs	microRNAs
CAPG	Macrophage-capping protein
NT	Nuclear transfer
OF	Oviductal fluids

OXTR	Oxytocin receptor
PDGF-AA	Platelet-derived growth factor-AA
PA	Parthenogenetic activation
PTr2	Porcine trophectoderm cells
PAOEC	Proliferating endothelial cells
siRNAs	Small RNAs
sEVs	Small extracellular vesicles
SCNT	Somatic cell nuclear transfer
tRFs	tRNA Glu-derived fragments
UF	Uterine fluid
ULF	Uterine luminal fluid
UFs	Uterine flushings
ULF-EVs	Uterine luminal fluid-derived extracellular vesicles
VEGF	Vascular endothelial growth factor

1 Introduction

1.1 Preimplantation Embryo Development

Implantation is crucial in establishing pregnancy in mammals, where embryos must adhere to and invade into the receptive endometrium. Successful implantation relies on the proper and coordinated preparation of the maternal endometrium and the embryo (Chen et al. 2022; Zhou and Dimitriadis 2020; Cha et al. 2012). Female gametes and early embryos interact with the surrounding environment in the female reproductive tract including the oviduct and uterine endometrium and have many opportunities to interact with extracellular vesicles (EVs) present there (Machtinger et al. 2016; Barkalina et al. 2015). Within the follicular fluid of the preovulatory follicle, EVs affect cumulus expansion and subsequent fertilization of the oocytes (Gabryś et al. 2024). Once released into the oviduct, the oocytes encounter sperm and possibly EVs that have traveled with the sperm, as well as EVs secreted by oviductal cells (Almiñana et al. 2017; Machtinger et al. 2016). Following fertilization within the ampulla of the oviduct, the cleaving embryo migrates down the entire length of the oviduct, with the surrounding cells and fluids in the microenvironment changing over time and distance (Swann 2014; Yuan et al. 2021).

The embryo's development is coordinated with its migration to the utero-tubal junction. As the embryo migrates, it undergoes cleavage from the single-cell zygote into two equal daughter cells, then four cells, and eight cells (Croxatto 2002; Taylor and Gomel 2008; Lonergan et al. 2016). At the 8–16 cell stage, depending on the species, the outer cells of the embryo undergo compaction (White et al. 2016). This makes the first stage of embryo differentiation, as the outer cells change shape, forming close cell-cell apposition with their neighboring cells; tight junctions to hold the cells together, gap junctions to provide open communication between cells, and the differentiation of the trophectoderm layer (White et al. 2016). The inner cells then become fully enclosed by the surrounding trophectoderm, and these cells enter differentiation pathways to become the inner cell mass and later the primitive endoderm (Copp 1979; Kim and Bedzhov 2022). During this stage of development, the trophoblast cells function as an epithelium and begin transporting water from the extracellular environment, secreting it to the inside of the embryo, thus producing and expanding the blastocoel cavity, resulting in the blastocyst stage of embryo development. It is at this stage that the embryo enters the uterus (Croxatto 2002; White et al. 2016).

During these early stages of development, as the embryo travels through the oviduct to the uterine horn, the cleavage stage embryo is enclosed within the zona pellucida which prevents direct cell-cell contact with the oviductal and uterine cells (Florman and Ducibella 2006; Kress and Selwood 2003). The zona may also limit the passage of EVs and other signaling molecules present in the surrounding microenvironment to the outer surface of the embryo (Munoz et al. 2022; Pavani et al. 2017). However, once it has entered the uterus, the blastocyst expands and hatches from the zona pellucida, either by rupture or by enzymatic digestion. Once the blastocysts are freed from the zona, the cells of the embryo come into direct contact with the components of uterine fluid and the surface epithelium of the endometrium. At this stage, both the blastocyst and the endometrial layer are likely releasing EVs into the intraluminal uterine environment, thus enabling an early mode of cross-talk between maternal and embryonic cells that contribute to successful implantation of the blastocyst (Das et al. 2023; Liang et al. 2017). Figure 1 provides a visual depiction of early embryo development and implantation.

1.2 Introduction to Extracellular Vesicles

The EVs are small, membrane-bound particles released by cells into the extracellular milieu and cannot replicate independently (Raposo and Stoorvogel 2013). They play a vital role in intercellular communication by transporting cargoes of various bioactive molecules (Welsh et al. 2024). These bioactive molecules include proteins, lipids, mRNAs, and microRNAs and can have a profound influence on the function and behavior of recipient target cells (Guo et al. 2023). The EVs include exosomes and microvesicles, differing in size and biogenesis. Exosomes, ranging from 20 to 150 nm in diameter, are produced inside cells through the inward budding of late endosomes, known as multi-vesicular bodies (Fig. 2). These exosomes are then released into the extracellular environment by fusing of the multi-vesicular body with the cell plasma membrane (Laulagnier et al. 2004; Modani et al. 2021). Microvesicles are formed by direct budding from the plasma membrane and typically have a diameter ranging from 50 to 1,000 nm (Traver et al. 2014; Crescitelli et al. 2013) (Fig. 2). Due to the overlapping size ranges, it is widely acknowledged that EV preparations are typically diverse, making it challenging to differentiate between the many subtypes of EVs (Jia et al. 2022). The term "exosomes" has been

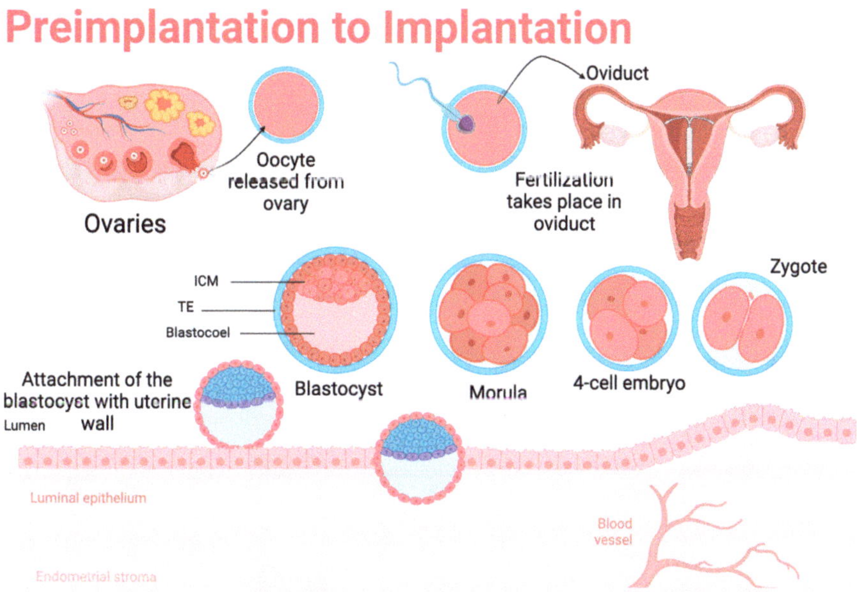

Fig. 1 The sequence of events from preimplantation to implantation of the embryo (BioRender.com)

widely used to refer to any vesicle discovered in a biofluid outside cells. However, the preferred nomenclature now is to use the term EVs (Welsh et al. 2024). EVs have also been categorized based on the specific tissue or biofluid in which they are identified. The terms prostatosomes or prostasomes, epididymosomes, oviductosomes, and uterosomes are used to refer to vesicles that are obtained from seminal fluid, epididymal fluid, oviductal fluid, and uterine fluid, respectively (Ronquist and Brody 1985; Saez et al. 2003; Griffiths et al. 2008; Al-Dossary et al. 2013). Surface molecules on EVs facilitate interactions with recipient target cells by adhering to lipids and ligands on the recipient cell's surface, leading to the internalization of the entire vesicle or the fusion of the EV membrane with the recipient cell's plasma membrane (Théry et al. 2009; Record et al. 2011). EVs are involved in various physiological and pathological processes, such as immune responses, cancer progression, cell metabolism, and tissue regeneration (Li et al. 2021; Zhang et al. 2016; Iraci et al. 2016; Lucchetti et al. 2020).

The EVs are secreted by many different types of cells in the body. Studies in female reproduction have focused on the impact of EVs in fluids such as ovarian follicular fluid, the oviduct, and the uterus, with EVs produced by cells of all these tissues as well as by the developing embryos (Hung et al. 2015; Almiñana and Bauersachs 2019; Burns et al. 2016). This review focuses on the EVs that provide communication between the embryo and endometrium, facilitating implantation.

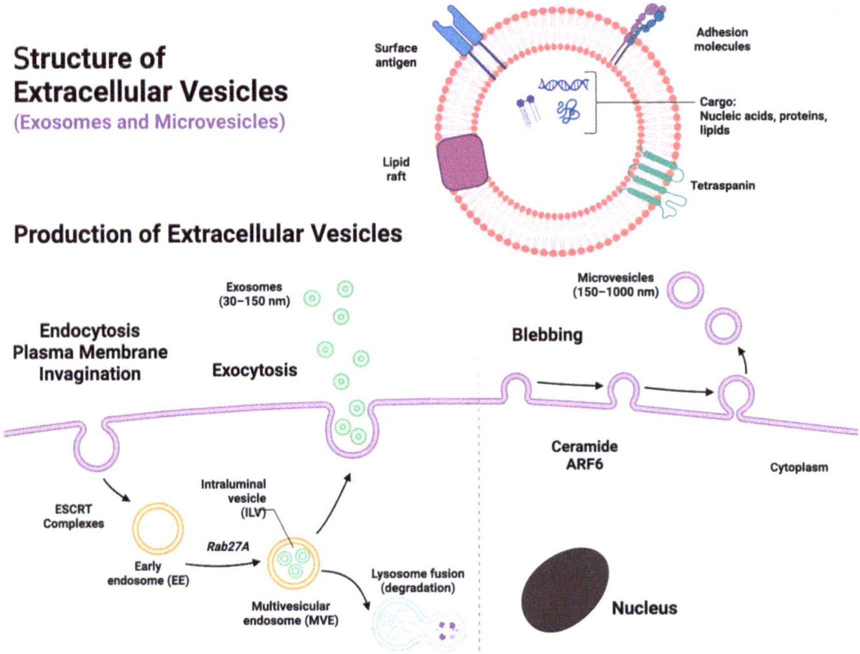

Fig. 2 Structure of extracellular vesicles and their biogenesis (BioRender.com)

1.3 EVs in the Female Reproductive Tract

The EVs have been isolated from female reproductive tract tissues and fluids. These EVs play an important role in cell-to-cell communication during early embryo development and implantation (Kurian and Modi 2019; Cajas et al. 2021; Hart et al. 2022). Proteomic and transcriptomic analyses of these EVs reveal that they carry different components that help in communication between the embryo and the maternal uterus (Muhandiram et al. 2023; Giacomini et al. 2021; Tan et al. 2020). Both the embryo and the uterine cells can take up EVs during early pregnancy (Aguilera et al. 2024; Burns et al. 2016) (Fig. 3). MicroRNAs (miRNAs) in these EVs play a role in the development of the embryo (Gross et al. 2017). In the uterus, EVs have been shown to play a role in uterine receptivity and to modulate the immune response during implantation (Hart et al. 2022; Giacomini et al. 2019). In the female reproductive tract, EVs have been isolated from follicular, oviductal, and uterine fluids and uterine cells (Burns et al. 2014; Gabryś et al. 2022; Harris et al. 2020).

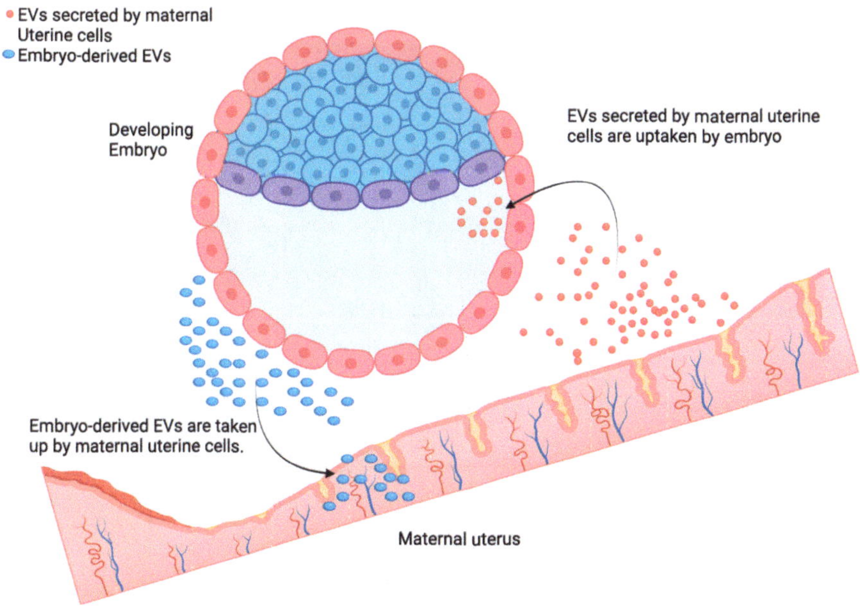

Fig. 3 Preimplantation embryo-maternal communication through EVs (BioRender.com)

2 Role of EVs in Embryo Implantation

Coordination between the maternal uterus and embryo is necessary for successful implantation (Zhou and Dimitriadis 2020). EVs from the uterus and the embryo play an important role in maternal-embryo cross-talk through the specific makeup of their cargos, which include proteins, lipids, metabolites, miRNAs, and nucleic acids (Liang et al. 2017). The EVs are secreted by both the developing embryo and maternal uterine cells. Embryo-derived EVs are taken up by endometrial cells and vice versa and play a role in embryo-maternal communication (Fig. 3).

2.1 The Role of EVs Present in Uterine Flushing on Embryo Development and Implantation

Different nutrients and soluble substances present in uterine fluid support the development and implantation of the embryo. These substances in the uterine fluid increase the expression of adhesion molecules in the endometrial epithelial cells and also promote embryo growth. EVs have been detected in uterine fluid (UFs) of different animals, including sheep, cattle, pigs, and humans (Ranjbaran et al. 2019; Nakamura et al. 2019, 2020; Hua et al. 2022; Xie et al. 2021; Capra and Lange-

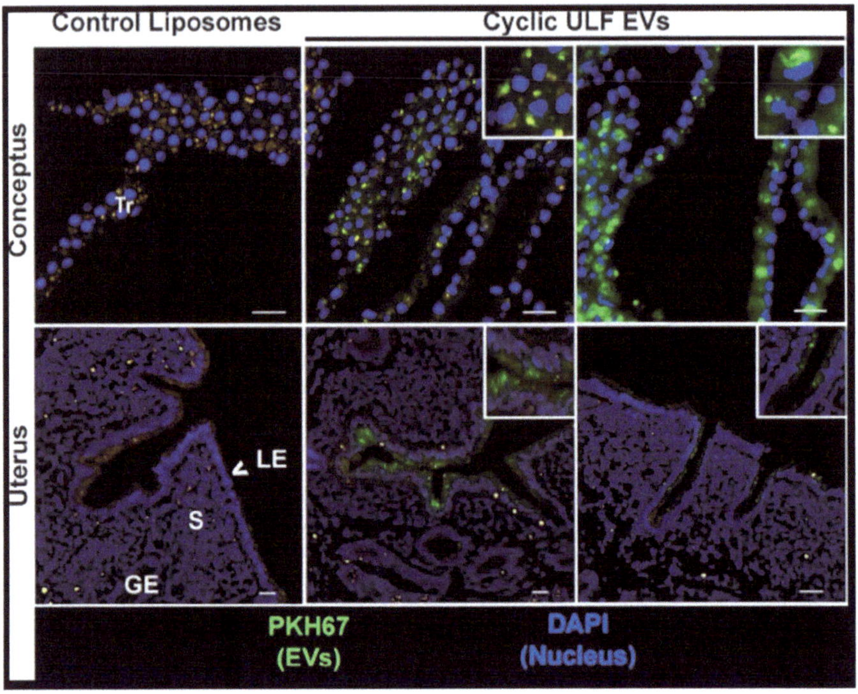

Fig. 4 Internalization of EVs in the conceptus trophectoderm and uterine epithelial cells (Reproduced with permission from Burns et al. 2016)

Consiglio 2020). These EVs are taken up by other neighboring uterine endometrial cells and by the embryo (Ma et al. 2022; Segura-Benítez et al. 2023; Pavani et al. 2022). Burns et al. showed the internalization of EVs present in uterine luminal fluid (UFL) by both uterine epithelial cells and the conceptus trophectoderm. UFL-EVs were isolated and labeled with PKH67 dye in day 14 cyclic sheep. Labeled UFL-EVs were then infused for 6 days into the uterine lumen of pregnant sheep starting at day 8 of pregnancy through an osmotic pump. These labeled ULF-EVs were found to be present in uterine epithelial cells and in conceptus trophectoderm on day 14 of pregnancy (Burns et al. 2016) (Fig. 4).

2.2 *The Role of EVs Present in Uterine Flushing on Maternal Uterine Receptivity*

Implantation is a complex process that requires careful coordination between the receptive uterus and the embryo (Zhou and Dimitriadis 2020; Chen et al. 2022). Uterine fluid contains numerous substances and molecules that help coordinate the proper timing of maternal endometrium receptivity. EVs derived from uterine flushings significantly influence the uterine environment and facilitate implantation (Beal et al. 2023). For example, EVs isolated from bovine uterine flushing on days 17, 20, and 22 of pregnancy were found to contain essential proteins, including interferon-tau (IFNT) (Kusama et al. 2018). These EVs were internalized by endometrial epithelial cells (EECs), where they upregulated the expression of interferon-stimulated genes (ISGs) such as ISG15, MX1, MX2, STAT1, and STAT2, which are critical for preparing a receptive uterine environment for implantation (Kusama et al. 2018). Additionally, EVs from day 17 of pregnancy upregulated apoptosis-related genes (BAX, CASP3, TNFA, TP53) in endometrial epithelial cells (EECs), facilitating the necessary remodeling of the endometrial lining for embryo attachment. EVs from days 20 and 22 of pregnancy further enhanced the expression of the adhesion molecule VCAM1, promoting stronger attachment of the conceptus to the uterine epithelium (Kusama ct al. 2018).

Global transcriptomic analysis of UF-EVs showed substantial transcriptomic alterations between the receptive phase (7 days post-LH surge) and the non-receptive phase (2 days post-LH surge) in women (Giacomini et al. 2021). The analysis identified 942 genes more abundantly expressed in UF-EVs during the receptive phase and 1,305 genes that were less abundantly expressed in UF-EVs during the receptive phase compared to the non-receptive phase. These differentially expressed genes are involved in cell adhesion, immune regulation, and response to hormonal stimuli, all essential for preparing the endometrium for successful embryo implantation. These findings suggest that UF-EVs may also serve as non-invasive biomarkers for real-time monitoring of endometrial receptivity (Giacomini et al. 2021).

MicroRNAs in uterine flushing (UF) EVs have been found to modulate the maternal immune response during the peri-implantation period. Nakamura et al. investigated how miR-98 present in UF-EVs regulates the maternal immune response during the peri-implantation period in cattle (Nakamura et al. 2019). Bovine endometrial epithelial cells (EECs) were treated with UF-EVs collected on days 17 (P17) and 20 (P20) of pregnancy. RNA sequencing analysis revealed that transcripts related to the immune system were significantly downregulated in EECs treated with EVs from P20 compared to those from P17. Through microRNA sequencing and in silico analysis, miR-98 was identified as a key miRNA in EVs that potentially regulates the maternal immune response (Nakamura et al. 2019). Synthetic miR-98 downregulated several immune system-related genes, such as CTSC, IL6, CASP4, and IKBKE in EECs. These findings suggest that miR-98 present in EVs contributes to promoting a uterine environment conducive to conceptus attachment and implantation by modulating maternal immune responses in the endometrium during the peri-implantation period to promote immune tolerance of the embryo (Nakamura et al. 2019).

Small RNAs present in EVs may play a role in endometrial receptivity and embryo development during the peri-implantation period in other species as well. EVs were isolated from uterine flushings obtained from gilts on days 10, 13, and 18 of gestation, and their RNA content was profiled using small RNA sequencing. The analysis identified 152 known miRNAs, 43 novel miRNAs, 6,248 known piRNAs, and 110 novel piRNAs present in the EVs (Hua et al. 2021). Differential expression of these small RNAs was observed across the different stages of gestation. Specifically, miRNAs including ssc-let-7f-5p, ssc-let-7i-5p, and ssc-let-7g showed significant changes in expression. Bioinformatics analysis revealed that the differentially expressed miRNAs are involved in processes crucial for the immune response, endometrial receptivity, and embryo development (Hua et al. 2021).

2.3 The Effect of EVs in Uterine Flushings on Embryo Development

The EVs that are present in the uterine fluid can be taken up by the embryo (Fig. 4). These EVs carry various cargos that also play an important role in maternal-to-embryo communication and embryo development. Hu et al. reported that EVs isolated from porcine uterine luminal fluid (ULF) were internalized by both trophoblast cells from normal embryos as well as parthenogenetic-activated embryos, where they modified the proliferation and migration of trophoblast cells (Hu et al. 2022). During the preimplantation period (before day 12) in pigs, conceptuses experience an extended phase of free migration within the uterus. As they reach day 12 of development, they become filamentous, their migration slows down, and they start to establish their final position within the uterus (Bazer and Johnson 2014; Hong et al. 2013). EVs from ULF on days 12 and 15 of pregnancy inhibited trophoblast cell proliferation and migration, suggesting a role in positioning of the conceptus for successful implantation. Small RNA sequencing identified 305 miRNAs within these EVs, with enrichment analyses indicating their involvement in molecular transport, signaling pathways, and immune regulation (Hu et al. 2022). Another study showed that oviductal and uterine fluid EVs significantly enhance bovine embryo quality when supplemented in a sequential, in vitro culture (IVC) medium. EVs significantly improved key quality parameters such as blastocyst survival rates post-vitrification, blastocyst total cell numbers, and lipid content (Leal et al. 2022).

Qiao et al. reported that UF-EVs improved the developmental competence of somatic cell nuclear transfer (SCNT) embryos (Qiao et al. 2018). EVs were successfully isolated from early luteal-phase bovine uterine fluid and supplemented into day 4 SCNT embryo cultures. The addition of the EVs significantly enhanced blastocyst formation and hatching rates compared to the control and supplementation performed on other days of culture. The day 4 supplementation group exhibited a higher inner cell mass/trophectoderm (ICM/TE) cell ratio and a lower apoptotic index, indicating improved embryo quality. Gene expression analysis revealed increased levels of IFNT and acrogranin and decreased levels of the stress-related genes HSP70, BAX, and BIP in the EV-supplemented embryos. These findings suggest that bovine uterus-derived EVs play a crucial role in enhancing the developmental competence and quality of SCNT embryos, potentially improving cloning efficiency and offering new insights into improving assisted reproductive technologies (Qiao et al. 2018).

Other recent studies have shown that uterine luminal fluid-derived extracellular vesicles (ULF-EVs) hold a significant potential as nanomaterials to improve embryo implantation. Proteomic analysis of ULF-EVs found that 1,549 proteins were enriched in these EVs. These proteins may play crucial roles in promoting embryo implantation by facilitating cell communication and cargo transfer. Specifically, proteins such as MEP1B, encapsulated in ULF-EVs, have been found to enhance trophoblast cell proliferation and migration, essential processes for successful implantation (Hong et al. 2023). Another study identified 333 miRNAs within bovine UF-EVs involved in post-transcriptional regulation, impacting key lipid metabolism and embryonic development genes (Mazzarella et al. 2024). EVs were isolated from the UF of heifers that were slaughtered between days 5 and 10 of the estrous cycle. These UF-EVs contained miRNAs including miR-155 and miR-24-3p, which are involved in cell differentiation, proliferation, and immune modulation. These findings highlight the critical role of EV-coupled miRNAs in regulating lipid metabolism and other essential pathways, contributing to improved embryo quality and successful pregnancy outcomes (Mazzarella et al. 2024).

3 Role of EVs Secreted by Uterine Cells in Embryo-Maternal Cross-Talk: In Vitro Studies

3.1 Proteins Identified in EVs from Uterine Cells

Proteomic analysis of EVs secreted by primary human epithelial endometrial cells and proliferating endothelial cells (PAOEC) in pigs identified 218 and 150 proteins, respectively, including integrins and the glycoprotein mucin, which are involved in biological processes such as cell adhesion, cell differentiation, migration, extracellular matrix organization, and vascular development (Segura-Benítez et al. 2022; Bidarimath et al. 2017). Ma et al. showed that EVs secreted by human uterine stromal cells regulate decidualization, angiogenesis, and trophoblast differentiation (Ma et al. 2022). Secretion of EVs during decidualization was regulated by the HIF2α-RAB27B pathway, and the EVs contain a variety of proteins that influence target cell functions. One mechanism by which these EVs promoted decidualization was by increasing glucose uptake through GLUT1, thus supporting the increased energy needs of stromal cells required for differentiation to decidual cells (Ma et al. 2022). The EVs also stimulated endothelial cell proliferation and vascular network formation, which are essential for maternal angiogenesis at the implantation site (Ma et al. 2022).

3.2 miRNAs Identified in EVs from Uterine Cells

EVs contain miRNAs involved in regulating angiogenesis, cell adhesion, and other physiological processes needed for the development of the placenta. PAOEC-EVs contain miRNAs involved in the regulation of angiogenesis (Bidarimath et al. 2017). Hunt Ng et al. reported that EVs from human endometrial epithelial (HEE) cells contained a distinct set of miRNAs, including hsa-miR-200c, hsa-miR-17, and hsa-miR-106a (Ng et al. 2013). In a recent study, Segura-Benitez identified 149 annotated miRNAs, with 37 identified as most relevant functionally for implantation. The target genes of these miRNAs are involved in processes essential for embryo development, cell cycle regulation, differentiation, apoptosis, metabolism, and gene expression. Key miRNAs such as hsa-miR-92a-3p, hsa-let-7b-5p, and hsa-miR-21-5p act as master regulators, enhancing endometrial receptivity and supporting early embryo development (Segura-Benítez et al. 2023). Tan et al. showed that human endometrial epithelial cells (EECs) secrete small extracellular vesicles (sEVs) enriched with miR-100-5p during the window of implantation, promoting angiogenesis and enhancing trophoblast functions such as migration, invasion, and proliferation in a human trophoblast cell line (Tan et al. 2021). MicroRNA-100-5p activates focal adhesion kinase (FAK) and c-Jun N-terminal kinase (JNK) pathways, facilitating the migration and invasion capabilities of trophoblast cells and enhancing the proliferation and migration of human umbilical vein endothelial cells (HUVECs). These findings highlight a novel mechanism of maternal-embryo communication through sEV miR-100-5p (Tan et al. 2021).

Wang et al. investigated how EVs derived from primary bovine endometrial epithelial cells may play an important role in supporting trophoblast cell development through the targeting of secreted frizzled-related protein 2 (sFRP2) by miR-218 (Wang et al. 2021). EVs isolated from epithelial cells contain miR-218, which is significantly downregulated in EECs of cows with endometritis. When miR-218 is taken up by placental trophoblast cells via EVs, it directly targets and downregulates sFRP2, a known inhibitor of the Wnt signaling pathway. By inhibiting sFRP2, miR-218 promotes the activation of Wnt signaling, which is essential for the migration, invasion, and overall developmental competence of trophoblast cells (Wang et al. 2021). A study by Evans et al. used an in vitro spheroid model of human trophectoderm, representing the preimplantation embryo, to investigate the interaction between these trophectoderm cells and endometrial-derived EVs. EVs isolated from the human endometrial adenocarcinoma cell line (EEC-1) that had been treated with estrogen and progesterone were internalized by the trophectoderm cells, leading to enhanced adhesive and invasive capacities of the trophectoderm spheroids (Evans et al. 2019). Proteomic analysis revealed significant alterations in the cellular proteome of trophectoderm spheroids following EV treatment, with upregulation of proteins involved in cell adhesion, cytoskeleton regulation, and metabolic processes (Evans et al. 2019).

Another recent study demonstrated how EVs from bovine endometrial cells and from bovine uterine fluid influence preimplantation embryo development through miRNAs. EVs from endometrial epithelial cells (EECs) and uterine fluid (UF-EVs) were introduced to in vitro-produced bovine embryos in culture. These EVs were internalized by the embryos leading to increased embryo diameter and a higher percentage of expanded and hatched blastocysts. Gene expression analysis revealed that embryos treated with UF-EVs showed increased interferon-tau (IFNT) expression, a marker of embryo quality and important for maternal recognition of pregnancy. Embryos treated with EVC EVs showed upregulation of pluripotency genes (Sox2 and Nanog) and a reduction in oxidative stress and apoptosis markers (SOD1 and BAX) (Aguilera et al. 2024).

3.3 Effect of EVs from Endometrial Mesenchymal Stromal Cells on Embryo Development

Endometrial mesenchymal stromal cells (endMSCs) are multipotent cells within the uterine lining that contribute to endometrial regeneration and remodeling (Khoury et al. 2014). These were isolated from menstrual blood of the first 3 days of the menstrual phase (Moreno et al. 2017; Zemelko et al. 2013). Exposure of murine embryos to EVs derived from the endMSCs led to significant developmental improvements in the cultured embryos (Blazquez et al. 2018). The authors demonstrated that the EV-endMSCs were able to enhance the release of vascular endothelial growth factor (VEGF) and platelet-derived growth factor-AA (PDGF-AA) from the embryos, indicating a dose-dependent pro-angiogenic effect. This release of pro-angiogenic factors was associated with an increase in the total blastomere count in the embryos as well as improved hatching rates of the murine blastocysts. The beneficial effects observed in this study suggest that EV-endMSCs can not only promote overall cell proliferation within the developing blastocysts but also likely support endometrial angiogenesis, vascularization, and tissue remodeling at the implantation site through modulation of gene expression in the embryos, all of which are critical for successful embryo development and implantation (Blazquez et al. 2018).

3.4 Autocrine Effects of EVs from Preimplantation Embryos

Since most cells and tissues produce EVs, it is not surprising that preimplantation embryos have also been shown to release EVs. Saadeldin et al. isolated EVs from the culture medium of in vitro cultured parthenogenetic porcine embryos. These EVs were found to contain mRNAs for key pluripotency genes (Oct4, Sox2, Klf4, c-Myc, and Nanog) that are expressed in preimplantation embryos. When these EVs were labeled with PKH67 dye and added to culture drops containing nuclear transfer porcine embryos, the EVs were internalized by the embryos and improved the in vitro survival, cleavage, and blastocyst formation rates when compared to embryos cultured without EVs. This was particularly interesting since mammalian embryos often display higher rates of development in vitro when cultured in groups versus single embryo culture. These results suggest that EVs likely serve as a novel means of embryo-to-embryo communication, offering a new paradigm for enhancing the development of cloned embryos through co-culture with other embryos (Saadeldin et al. 2014).

Qu et al. reported that embryo-derived EVs significantly enhanced the development of cloned bovine embryos. Bovine somatic cell nuclear transfer embryos cultured in a nonrenewal system, which retains all EVs in the medium, showed a higher blastocyst formation rate (37.8% vs. 23.1%), increased total cell numbers (108.0 vs. 91.9), and a lower apoptotic index (1.9% vs. 6.3%) compared to the renewal system where medium is refreshed and EVs removed. The nonrenewal group also had a better ICM/TE ratio (41.9% vs. 30.3%) and higher expression of the pluripotency marker Oct-4 and anti-apoptosis marker Bcl-2, while the pro-apoptosis marker Bax and stress marker Bip showed lower expression. Supplementation of EVs into a renewed medium partially rescued the impaired development of these embryos, improving blastocyst formation to 30.8%, increasing total cell numbers (99.1) and ICM/TE ratio (39.5%), and enhancing Oct-4 expression. Embryos from the nonrenewal and EV-supplemented groups ultimately had higher pregnancy rates and calving rates, comparable to each other, indicating that EVs play an important role in supporting both in vitro and in vivo development of SCNT embryos (Qu et al. 2017).

3.5 *Embryo-Derived EVs as Markers of Embryo Quality*

Good-quality embryos are necessary for successful implantation and pregnancy outcomes, particularly in assisted reproductive technologies. EVs secreted by embryos can be used as a non-invasive means to evaluate embryo quality. Mellisho et al. found that in vitro fertilization (IVF)-produced bovine embryos secrete more EVs as compared to parthenogenetic activation (PA) embryos (Mellisho et al. 2017). They isolated EVs from days 7 to 9 in cultured embryos and then observed embryo growth until day 11. The number of EVs secreted by the two types of blastocysts was different. IVF-derived blastocysts were observed to secrete a higher number of EVs compared to PA-derived blastocysts. Non-competent blastocysts (those with arrested development at day 11) secreted a greater number of EVs than their competent counterparts, particularly in IVF-derived embryos (Mellisho et al. 2017). In a separate study, Melo Baez et al. reported that the EVs secreted during the compaction stage (days 3.5–5) and blastulation stage (days 5–7) contained distinct miRNA profiles, with 79 miRNAs differentially expressed between these stages (Melo-Baez et al. 2020). The miRNAs in embryo-derived EVs such as bta-miR-103, bta-miR-100, bta-miR-502a, and bta-miR-1 were upregulated in EVs from less competent embryos, whereas bta-miR-92a, bta-miR-140, bta-miR-2285a, and bta-miR-222 were downregulated (Melo-Baez et al. 2020). The miRNAs identified regulate various signaling pathways essential for embryo development, such as TGF-β, Ras, and Wnt signaling pathways. Therefore, miRNAs and the number of EVs in the culture medium could serve as non-invasive biomarkers for assessing embryo quality, helping to identify embryos with the highest potential for successful development and implantation (Melo-Báez et al. 2023; Mellisho et al. 2017; Dissanayake et al. 2020).

Fan et al. investigated the role of tRNA Glu-derived fragments (tRFs) from embryo-derived EVs in modulating bovine embryo hatching (Fan et al. 2024). Small RNA sequencing revealed that these tRFs, specifically tDR-14:32-Glu-CTC-1, were significantly higher in EVs from pre-blastocyst embryos compared to blastocyst-stage embryos. Functional assays using antagomirs to inhibit tDR-14:32-Glu-CTC-1 demonstrated that its suppression improved embryo hatching rates. The transcriptomic analysis further showed that inhibiting tDR-14:32-Glu-CTC-1 altered the expression of genes associated with actin cytoskeleton formation and embryo attachment, such as ACTA1, ACTA2, LCP1, and ATF3, thereby enhancing structural support for embryo hatching (Fan et al. 2024).

Embryo-derived EVs contain specific microRNAs that promote blastocyst hatching. Pavani et al. reported that the hatching process of bovine blastocysts was modulated by microRNA-378a-3p, which is present in EVs secreted by the blastocysts. Analysis of the EVs isolated from the culture medium of individual bovine embryos identified 69 miRNAs that were differentially expressed between embryos that developed into blastocysts and those that did not. Among these, miR-378a-3p was specifically upregulated in the EVs from blastocysts. Functional assays using miR-378a-3p mimics and inhibitors revealed that supplementation with the miR-378a-3p mimics significantly improved blastocyst quality by increasing cell numbers, reducing apoptosis, and enhancing hatching rates. The miR-378a-3p inhibitor reduced hatching rates and embryo quality. Transcriptomic analysis showed that miR-378a-3p modulates the expression of genes associated with embryo development and implantation. These findings suggest that miR-378a-3p in EVs may play an important role in regulating blastocyst hatching and can be a valuable non-invasive marker for assessing embryo quality (Pavani et al. 2022).

3.6 *Effect of Embryo-Derived EVs on Maternal Endometrium*

EVs secreted by the embryo also play an important role by directly coordinating the preparation of the maternal endometrium. EVs from both in vitro-produced (IVP) and in vivo-produced (IVV) embryos are internalized by bovine endometrial epithelial cells (bEECs). The EVs from IVP embryos increased the expression of the oxytocin receptor (OXTR) and non-classical ISGs such as CST6 and WNT7A (Aguilera et al. 2024). EVs secreted by pre-hatching bovine embryos, whether produced in vitro or in vivo, significantly altered the expression of IFNtau-stimulated genes such as OAS1Y, MX1, MX2, and ISG15, as well as other genes related to endometrial function, including PTGS1, PTGS2, and MUC1 in bovine endometrial cells (Aguilera et al. 2023). The EVs from IVP embryos had a more pronounced effect, altering the expression of 3,552 genes compared to 1,839 genes for EVs from IVV embryos (Aguilera et al. 2023).

Nakamura et al. investigated the role of conceptus-derived exosomes in regulating the endometrial response during the attachment period in a sheep model. The EVs were isolated from UF of day 15 cyclic (C15), day 17 cyclic (C17), day 15 pregnant (P15), and day 17 pregnant (P17) ewes. The EVs from UF of P15 and P17 ewes contained IFNT, macrophage-capping protein (CAPG), and aldo-keto reductase family 1 member B1 (AKR1B1), which were absent in EVs from cyclic ewes. The EVs from UF of pregnant ewes upregulated the expression of ISGs such as STAT1, STAT2, MX1, MX2, BST2, and ISG15 in EECs (Nakamura et al. 2016). Poh et al. demonstrated that EVs secreted by human trophectoderm cells are enriched with critical players involved in implantation, including proteins that regulate antioxidant activity (SOD1, PRDX1/4/6), tissue integrity (EZR, RAC1, RHOA, TNC), and focal adhesions (FAK, ITGA2/V, ITGB1/3) (Poh et al. 2021). All of these studies support that conceptus-derived EVs facilitate conceptus-endometrium interaction by promoting a uterine environment conducive to implantation (Nakamura et al. 2016; Poh et al. 2021).

EVs secreted by human aneuploid embryos present a distinct transcriptomic profile and significantly impact the expression of genes in decidualizing endometrial stromal cells (dESCs) (Makieva et al. 2024). The EVs from aneuploid embryos contain specific RNA transcripts, such as PPM1J, LINC00561, ANKRD34C, and TMED10, which differ in abundance from those present in EVs of euploid embryos. The EVs from aneuploid embryos induced upregulation of the MUC1 gene in dESCs: a gene typically downregulated during decidualization to facilitate endometrial receptivity for embryo implantation. This increase in MUC1 expression, a glycoprotein that acts as a physical barrier to embryo attachment, suggests a mechanism by which aneuploid embryos might signal their developmental incompetence, leading to implantation failure. This example of a distinct transcriptomic response highlights the potential of using EV-derived miRNAs as non-invasive biomarkers for embryo ploidy status and provides insights into the molecular interactions governing embryo-endometrial communication and implantation (Makieva et al. 2024).

Embryo-derived EVs have also been shown to impact the local intrauterine immune environment. Kusama et al. investigated the impact of day 7 embryos on the proteome and EV miRNAs' content of bovine uterine fluid and their potential role in innate immune functions (Kusama et al. 2021). The UF was collected from superovulated cows through an embryo transfer program consisting of either artificially inseminated (AI) or non-inseminated (control) cows. Results indicated that the presence of multiple day 7 embryos in the uterus significantly altered the protein composition and miRNA content of the UF compared to UF from non-inseminated cows. Specifically, the study identified 336 proteins in the UF, with 260 proteins showing over a twofold increase in inseminated cows that had been inseminated. Many of these proteins were associated with neutrophil-mediated immune responses suggesting an effect on the innate immune environment. EV miRNAs were also differentially expressed, with miR-29a, miR-199b, and miR-423-5p being notably linked to regulation of several uterine proteins. These changes in protein and miRNA profiles underscore the critical role of day 7 embryos in modulating the uterine environment to promote embryo tolerance and the establishment of pregnancy through innate immune functions (Kusama et al. 2021).

Burns et al. investigated how EVs derived from elongated ovine embryos may modulate the uterine environment and support embryo implantation (Burns et al. 2016). The EVs were isolated from conditioned medium of 24 h ex vivo cultured day 14 ovine conceptuses. Proteomic analysis of conceptus-derived EVs revealed the presence of 231 proteins that were enriched for specific protein classes, including chaperones, chaperonins, proteases, and protease inhibitors. RNA sequencing of conceptus-derived EVs identified 512 miRNAs. EVs were labeled with PKH67 dye and infused into the uterine lumen of cycling sheep, and they were observed in uterine epithelial cells at day 14 (Fig. 4) (Burns et al. 2016). The EVs isolated from porcine trophectoderm (PTr2) cells were shown to increase the proliferation of PAOEC cells. RT-PCR of PTr2-EVs identified the presence of 14 miRNAs, which are involved in regulation of angiogenesis and placental development. Proteomic analysis of PTr2-EVs identified 187 proteins involved in different biological processes (Bidarimath et al. 2017).

3.7 The Effect of EVs Derived from Embryonic Stem Cell Lines on Maternal Endometrial Receptivity

There is interest in using embryonic stem cell lines as a source of EVs for several purposes. EVs shed by embryonic stem (ES) cells have been shown to enhance trophoblast migration during implantation by carrying extracellular matrix proteins such as laminin and fibronectin (Desrochers et al. 2016). These proteins interact with integrins on the trophoblast cell surface, activating signaling pathways that promote migration – a key step in implantation (Desrochers et al. 2016). The EVs derived from placental trophoblast cells, particularly under the influence of progesterone, enhance endometrial epithelial receptivity by upregulating factors such as integrins αv and $\beta 3$ while downregulating anti-adhesion factors such as MUC1 (Su et al. 2022). The proteome of these EVs modulates the extracellular environment and cell metabolism, further promoting an environment conducive to implantation. These findings suggest that EVs from embryonic stem cell lines can play a role in improving embryo-maternal communication, thereby enhancing the likelihood of successful implantation and pregnancy establishment (Desrochers et al. 2016; Su et al. 2022).

3.8 Impact of EVs Derived from Cows with Endometritis on Embryo Development

Wang et al. investigated the impact of endometrial EV-derived microRNAs isolated from cows with endometritis on blastocyst development (Wang et al. 2019). Their results revealed that uterine EVs from cows with endometritis significantly impaired the development of IVF embryos compared to uterine EVs isolated from healthy cows. Specifically, supplementation of embryo cultures with EVs from cows with endometritis decreased the blastocyst formation rate from 32.37% to 21.84% ($P < 0.05$). Subsequent analysis identified 118 differentially expressed miRNAs in the EVs, with 66 upregulated and 52 downregulated in EVS from cows with endometritis. These differentially expressed miRNAs, validated by quantitative RT-PCR, suggest a mechanism by which the uterine endometrium of cows with a local uterine infection is functionally altered, potentially contributing to the lower fertility observed in cows with endometritis by affecting key developmental pathways (Wang et al. 2019).

4 Conclusion

EVs contain a multitude of different cargos including proteins, lipids, miRNAs, and other small RNAs that play a pivotal role in embryo development and implantation, serving as essential mediators of communication between the embryo and the maternal endometrium. Preimplantation embryo interacts with EVs secreted by both the oviduct and uterus. EVs are secreted by endometrial cells and endMSCs into the uterine fluid and are taken up by developing embryos. The ICM and TE cells of the embryo also secrete EVs that play a significant role in embryo-maternal communication. The evidence discussed in this review highlights that EVs, through their specific cargos of proteins, lipids, and nucleic acids, play a vital role in modulating key physiological processes including endometrial receptivity, immunomodulation, immune tolerance, embryo development, and the establishment of a conducive microenvironment for embryo attachment and invasion of trophoblast. Additionally, the differential expression of specific EV cargos during the implantation window suggests a highly regulated mechanism through which EVs contribute to implantation success or failure.

The findings discussed in this chapter not only advance our understanding of the role of EVs in embryo-maternal interaction but also open new avenues for therapeutic interventions in reproductive medicine. Targeting EV-mediated pathways could offer novel strategies for enhancing implantation rates in assisted reproductive technologies (ART) or addressing implantation failures in infertility treatments. Despite the significant advancements in understanding the role of EVs in implantation, several questions remain unanswered, providing fertile ground for future research. Key areas of interest include the precise molecular mechanisms through which EVs exert their influence on pathways involved in implantation, the identification of specific EV-associated biomarkers that could predict implantation success or failure, and the potential of EVs as therapeutic agents in the treatment of implantation-related disorders.

Acknowledgment

We gratefully acknowledge Keely Mogge for her significant contribution in sourcing and providing key articles that greatly supported the completion of this review.

Compliance with Ethical Standards This work was funded by the College of ACES FIRE grant at the University of Illinois. Authors Muhammad Ashir Nabeel and Romana Nowak declare that they have no conflict of interest. This chapter does not contain any studies with human participants or animals performed by any of the authors.

References

Aguilera C, Velásquez AE, Gutierrez-Reinoso MA, Wong YS, Melo-Baez B, Cabezas J, Caamaño D, Navarrete F, Rojas D, Riadi G (2023) Extracellular vesicles secreted by pre-hatching bovine embryos produced in vitro and in vivo Alter the expression of IFNtau-stimulated genes in bovine endometrial cells. Int J Mol Sci 24(8):7438

Aguilera C, Wong Y, Gutierrez-Reinoso M, Velásquez A, Melo-Báez B, Cabezas J, Caamaño D, Navarrete F, Castro F, Rodriguez-Alvarez L (2024) Embryo-maternal communication mediated by extracellular vesicles in the early stages of embryonic development is modified by in vitro conditions. Theriogenology 214:43–56

Al-Dossary AA, Strehler EE, Martin-DeLeon PA (2013) Expression and secretion of plasma membrane Ca2+-ATPase 4a (PMCA4a) during murine estrus: association with oviductal exosomes and uptake in sperm. PLoS One 8(11):e80181

Almiñana C, Bauersachs S (2019) Extracellular vesicles in the oviduct: progress, challenges and implications for the reproductive success. Bioengineering 6(2):32

Almiñana C, Corbin E, Tsikis G, Alcântara-Neto AS, Labas V, Reynaud K, Galio L, Uzbekov R, Garanina AS, Druart X (2017) Oviduct extracellular vesicles protein content and their role during oviduct–embryo cross-talk. Reproduction 154(3):253–268

Barkalina N, Jones C, Wood MJ, Coward K (2015) Extracellular vesicle-mediated delivery of molecular compounds into gametes and embryos: learning from nature. Hum Reprod Update 21(5):627–639

Bazer FW, Johnson GA (2014) Pig blastocyst–uterine interactions. Differentiation 87(1–2):52–65

Beal JR, Ma Q, Bagchi IC, Bagchi MK (2023) Role of endometrial extracellular vesicles in mediating cell-to-cell communication in the uterus: a review. Cells 12(22):2584

Bidarimath M, Khalaj K, Kridli RT, Kan FW, Koti M, Tayade C (2017) Extracellular vesicle mediated intercellular communication at the porcine maternal-fetal interface: a new paradigm for conceptus-endometrial cross-talk. Sci Rep 7(1):40476

Blazquez R, Sánchez-Margallo FM, Alvarez V, Matilla E, Hernandez N, Marinaro F, Gomez-Serrano M, Jorge I, Casado JG, Macias-Garcia B (2018) Murine embryos exposed to human endometrial MSCs-derived extracellular vesicles exhibit higher VEGF/PDGF AA release, increased blastomere count and hatching rates. PLoS One 13(4):e0196080

Burns G, Brooks K, Wildung M, Navakanitworakul R, Christenson LK, Spencer TE (2014) Extracellular vesicles in luminal fluid of the ovine uterus. PLoS One 9(3):e90913

Burns GW, Brooks KE, Spencer TE (2016) Extracellular vesicles originate from the conceptus and uterus during early pregnancy in sheep. Biol Reprod 94(3):56

Cajas YN, Cañón-Beltrán K, de la Blanca MGM, Sánchez JM, Fernandez-Fuertes B, González EM, Rizos D (2021) Role of reproductive fluids and extracellular vesicles in embryo–maternal interaction during early pregnancy in cattle. Reprod Fertil Dev 34(2):117–138

Capra E, Lange-Consiglio A (2020) The biological function of extracellular vesicles during fertilization, early embryo – maternal crosstalk and their involvement in reproduction: review and overview. Biomol Ther 10(11):1510

Cha J, Sun X, Dey SK (2012) Mechanisms of implantation: strategies for successful pregnancy. Nat Med 18(12):1754–1767

Chen K, Liang J, Qin T, Zhang Y, Chen X, Wang Z (2022) The role of extracellular vesicles in embryo implantation. Front Endocrinol 13:809596

Copp A (1979) Interaction between inner cell mass and trophectoderm of the mouse blastocyst: II. The fate of the polar trophectoderm. Development 51(1):109–120

Crescitelli R, Lässer C, Szabó TG, Kittel A, Eldh M, Dianzani I, Buzás EI, Lötvall J (2013) Distinct RNA profiles in subpopulations of extracellular vesicles: apoptotic bodies, microvesicles and exosomes. J Extracell Vesicles 2(1):20677

Croxatto HB (2002) Physiology of gamete and embryo transport through the fallopian tube. Reprod Biomed Online 4(2):160–169

Das PK, Mukherjee J, Banerjee D (2023) Fertilisation, gestation and parturition. In: Textbook of veterinary physiology. Springer, pp 569–613

Desrochers LM, Bordeleau F, Reinhart-King CA, Cerione RA, Antonyak MA (2016) Microvesicles provide a mechanism for intercellular communication by embryonic stem cells during embryo implantation. Nat Commun 7(1):11958

Dissanayake K, Nomm M, Lättekivi F, Ressaissi Y, Godakumara K, Lavrits A, Midekessa G, Viil J, Bæk R, Jørgensen MM (2020) Individually cultured bovine embryos produce extracellular vesicles that have the potential to be used as non-invasive embryo quality markers. Theriogenology 149:104–116

Evans J, Rai A, Nguyen HP, Poh QH, Elglass K, Simpson RJ, Salamonsen LA, Greening DW (2019) Human endometrial extracellular vesicles functionally prepare human trophectoderm model for implantation: understanding bidirectional maternal-embryo communication. Proteomics 19(23):1800423

Fan Y, Pavani KC, Smits K, Van Soom A, Peelman L (2024) tRNAGlu-derived fragments from embryonic extracellular vesicles modulate bovine embryo hatching. J Anim Sci Biotechnol 15(1):23

Florman HM, Ducibella T (2006) Fertilization in mammals. Knobil Neill's Physiol Reprod 1:149–196

Gabryś J, Kij-Mitka B, Sawicki S, Kochan J, Nowak A, Łojko J, Karnas E, Bugno-Poniewierska M (2022) Extracellular vesicles from follicular fluid may improve the nuclear maturation rate of in vitro matured mare oocytes. Theriogenology 188:116–124

Gabryś J, Gurgul A, Szmatoła T, Kij-Mitka B, Andronowska A, Karnas E, Kucharski M, Wojciechowska-Puchałka J, Kochan J, Bugno-Poniewierska M (2024) Follicular fluid-derived extracellular vesicles influence on in vitro maturation of equine oocyte: impact on cumulus cell viability, expansion and transcriptome. Int J Mol Sci 25(6):3262

Giacomini E, Alleva E, Fornelli G, Quartucci A, Privitera L, Vanni V, Viganò P (2019) Embryonic extracellular vesicles as informers to the immune cells at the maternal–fetal interface. Clin Exp Immunol 198(1):15–23

Giacomini E, Scotti G, Vanni V, Lazarevic D, Makieva S, Privitera L, Signorelli S, Cantone L, Bollati V, Murdica V (2021) Global transcriptomic changes occur in uterine fluid-derived extracellular vesicles during the endometrial window for embryo implantation. Hum Reprod 36(8):2249–2274

Griffiths GS, Galileo DS, Reese K, Martin-DeLeon PA (2008) Investigating the role of murine epididymosomes and uterosomes in GPI-linked protein transfer to sperm using SPAM1 as a model. Mol Reprod Dev 75(11):1627–1636

Gross N, Kropp J, Khatib H (2017) MicroRNA signaling in embryo development. Biology 6(3):34

Guo X-R, Ma Y, Ma Z-M, Dai T-S, Wei S-H, Chu Y-K, Dan X-G (2023) Exosomes: the role in mammalian reproductive regulation and pregnancy-related diseases. Front Physiol 14:1056905

Harris EA, Stephens KK, Winuthayanon W (2020) Extracellular vesicles and the oviduct function. Int J Mol Sci 21(21):8280

Hart AR, Khan NLA, Godakumara K, Dissanayake K, Piibor J, Muhandiram S, Eapen S, Heath PR, Fazeli A (2022) The role of extracellular vesicles in endometrial receptivity and their potential in reproductive therapeutics and diagnosis. Reprod Biol 22(2):100645

Hong L, Hou C, Li X, Li C, Yu M (2013) Expression pattern of CD 34 at the Maternal–Foetal interface during pregnancy in pigs. Reprod Domest Anim 48(5):762–767

Hong L, Zang X, Hu Q, He Y, Xu Z, Xie Y, Gu T, Yang H, Yang J, Shi J (2023) Uterine luminal-derived extracellular vesicles: potential nanomaterials to improve embryo implantation. J Nanobiotechnol 21(1):79

Hu Q, Zang X, Ding Y, Gu T, Shi J, Li Z, Cai G, Liu D, Wu Z, Hong L (2022) Porcine uterine luminal fluid-derived extracellular vesicles improve conceptus-endometrial interaction during implantation. Theriogenology 178:8–17

Hua R, Wang Y, Lian W, Li W, Xi Y, Xue S, Kang T, Lei M (2021) Small RNA-seq analysis of extracellular vesicles from porcine uterine flushing fluids during peri-implantation. Gene 766: 145117

Hua R, Liu Q, Lian W, Gao D, Huang C, Lei M (2022) Transcriptome regulation of extracellular vesicles derived from porcine uterine flushing fluids during peri-implantation on endometrial epithelial cells and embryonic trophoblast cells. Gene 822:146337

Hung W-T, Hong X, Christenson LK, McGinnis LK (2015) Extracellular vesicles from bovine follicular fluid support cumulus expansion. Biol Reprod 93(5):117. 111–119

Iraci N, Leonardi T, Gessler F, Vega B, Pluchino S (2016) Focus on extracellular vesicles: physiological role and signalling properties of extracellular membrane vesicles. Int J Mol Sci 17(2):171

Jia Y, Yu L, Ma T, Xu W, Qian H, Sun Y, Shi H (2022) Small extracellular vesicles isolation and separation: current techniques, pending questions and clinical applications. Theranostics 12(15): 6548

Khoury M, Alcayaga-Miranda F, Illanes SE, Figueroa FE (2014) The promising potential of menstrual stem cells for antenatal diagnosis and cell therapy. Front Immunol 5:205

Kim YS, Bedzhov I (2022) Mechanisms of formation and functions of the early embryonic cavities. In: Seminars in cell & developmental biology. Elsevier, pp 110–116

Kress A, Selwood L (2003) Conceptus polarity and cell-zona adhesion during early cleavage (fertilized tubal egg to 8-cell stage) in the marsupial Sminthopsis macroura. Cells Tissues Organs 175(1):9–24

Kurian NK, Modi D (2019) Extracellular vesicle mediated embryo-endometrial cross talk during implantation and in pregnancy. J Assist Reprod Genet 36:189–198

Kusama K, Nakamura K, Bai R, Nagaoka K, Sakurai T, Imakawa K (2018) Intrauterine exosomes are required for bovine conceptus implantation. Biochem Biophys Res Commun 495(1): 1370–1375

Kusama K, Rashid MB, Kowsar R, Marey MA, Talukder AK, Nagaoka K, Shimada M, Khatib H, Imakawa K, Miyamoto A (2021) Day 7 embryos change the proteomics and exosomal micro-rnas content of bovine uterine fluid: involvement of innate immune functions. Front Genet 12: 676791

Laulagnier K, Motta C, Hamdi S, Roy S, Fauvelle F, Pageaux J-F, Kobayashi T, Salles J-P, Perret B, Bonnerot C (2004) Mast cell-and dendritic cell-derived exosomes display a specific lipid composition and an unusual membrane organization. Biochem J 380(1):161–171

Leal CLV, Cañón-Beltrán K, Cajas YN, Hamdi M, Yaryes A, Millán de la Blanca MG, Beltrán-Breña P, Mazzarella R, Da Silveira JC, Gutiérrez-Adán A (2022) Extracellular vesicles from oviductal and uterine fluids supplementation in sequential in vitro culture improves bovine embryo quality. J Anim Sci Biotechnol 13(1):116

Li SR, Man QW, Gao X, Lin H, Wang J, Su FC, Wang HQ, Bu LL, Liu B, Chen G (2021) Tissue-derived extracellular vesicles in cancers and non-cancer diseases: present and future. J Extracell Vesicles 10(14):e12175

Liang J, Wang S, Wang Z (2017) Role of microRNAs in embryo implantation. Reprod Biol Endocrinol 15:1–11

Lonergan P, Fair T, Forde N, Rizos D (2016) Embryo development in dairy cattle. Theriogenology 86(1):270–277

Lucchetti D, Ricciardi Tenore C, Colella F, Sgambato A (2020) Extracellular vesicles and cancer: a focus on metabolism, cytokines, and immunity. Cancers 12(1):171

Ma Q, Beal JR, Bhurke A, Kannan A, Yu J, Taylor RN, Bagchi IC, Bagchi MK (2022) Extracellular vesicles secreted by human uterine stromal cells regulate decidualization, angiogenesis, and trophoblast differentiation. Proc Natl Acad Sci 119(38):e2200252119

Machtinger R, Laurent LC, Baccarelli AA (2016) Extracellular vesicles: roles in gamete maturation, fertilization and embryo implantation. Hum Reprod Update 22(2):182–193

Makieva S, Giacomini E, Scotti GM, Lazarevic D, Pavone V, Ottolina J, Bartiromo L, Schimberni M, Morelli M, Alteri A (2024) Extracellular vesicles secreted by human aneuploid embryos present a distinct transcriptomic profile and upregulate MUC1 transcription in decidualised endometrial stromal cells. Hum Reprod Open 2024(2):hoae014

Mazzarella R, Cañón-Beltrán K, Cajas YN, Hamdi M, González EM, da Silveira JC, Leal CL, Rizos D (2024) Extracellular vesicles-coupled miRNAs from oviduct and uterus modulate signaling pathways related to lipid metabolism and bovine early embryo development. J Anim Sci Biotechnol 15(1):51

Mellisho EA, Velásquez AE, Nuñez MJ, Cabezas JG, Cueto JA, Fader C, Castro FO, Rodríguez-Álvarez L (2017) Identification and characteristics of extracellular vesicles from bovine blastocysts produced in vitro. PLoS One 12(5):e0178306

Melo-Baez B, Wong YS, Aguilera CJ, Cabezas J, Mançanares AC, Riadi G, Castro FO, Rodriguez-Alvarez L (2020) MicroRNAs from extracellular vesicles secreted by bovine embryos as early biomarkers of developmental competence. Int J Mol Sci 21(23):8888

Melo-Báez B, Mellisho EA, Wong YS, Cabezas J, Caamaño D, Aguilera C, Riadi G, Castro FO, Rodriguez-Alvarez L (2023) Bovine embryos release extracellular vesicles with differential miRNA signature during the compaction and blastulation stages. Reprod Biol 23(1):100725

Modani S, Tomar D, Tangirala S, Sriram A, Mehra NK, Kumar R, Khatri DK, Singh PK (2021) An updated review on exosomes: biosynthesis to clinical applications. J Drug Target 29(9):925–940

Moreno R, Rojas L, Villellas FV, Soriano VC, García-Castro J, Fajardo C, Alemany R (2017) Human menstrual blood-derived mesenchymal stem cells as potential cell carriers for oncolytic adenovirus. Stem Cells Int 1:3615729

Muhandiram S, Dissanayake K, Orro T, Godakumara K, Kodithuwakku S, Fazeli A (2023) Secretory proteomic responses of endometrial epithelial cells to trophoblast-derived extracellular vesicles. Int J Mol Sci 24(15):11924

Munoz EL, Fuentes FB, Felmer RN, Yeste M, Arias ME (2022) Extracellular vesicles in mammalian reproduction: a review. Zygote 30(4):440–463

Nakamura K, Kusama K, Bai R, Sakurai T, Isuzugawa K, Godkin JD, Suda Y, Imakawa K (2016) Induction of IFNT-stimulated genes by conceptus-derived exosomes during the attachment period. PLoS One 11(6):e0158278

Nakamura K, Kusama K, Ideta A, Kimura K, Hori M, Imakawa K (2019) Effects of miR-98 in intrauterine extracellular vesicles on maternal immune regulation during the peri-implantation period in cattle. Sci Rep 9(1):20330

Nakamura K, Kusama K, Suda Y, Fujiwara H, Hori M, Imakawa K (2020) Emerging role of extracellular vesicles in embryo–maternal communication throughout implantation processes. Int J Mol Sci 21(15):5523
Ng YH, Rome S, Jalabert A, Forterre A, Singh H, Hincks CL, Salamonsen LA (2013) Endometrial exosomes/microvesicles in the uterine microenvironment: a new paradigm for embryo-endometrial cross talk at implantation. PLoS One 8(3):e58502
Pavani KC, Alminana C, Wydooghe E, Catteeuw M, Ramírez MA, Mermillod P, Rizos D, Van Soom A (2017) Emerging role of extracellular vesicles in communication of preimplantation embryos in vitro. Reprod Fertil Dev 29(1):66–83
Pavani KC, Meese T, Pascottini OB, Guan X, Lin X, Peelman L, Hamacher J, Van Nieuwerburgh F, Deforce D, Boel A (2022) Hatching is modulated by microRNA-378a-3p derived from extracellular vesicles secreted by blastocysts. Proc Natl Acad Sci 119(12):e2122708119
Poh QH, Rai A, Carmichael II, Salamonsen LA, Greening DW (2021) Proteome reprogramming of endometrial epithelial cells by human trophectodermal small extracellular vesicles reveals key insights into embryo implantation. Proteomics 21(13–14):2000210
Qiao F, Ge H, Ma X, Zhang Y, Zuo Z, Wang M, Zhang Y, Wang Y (2018) Bovine uterus-derived exosomes improve developmental competence of somatic cell nuclear transfer embryos. Theriogenology 114:199–205
Qu P, Qing S, Liu R, Qin H, Wang W, Qiao F, Ge H, Liu J, Zhang Y, Cui W (2017) Effects of embryo-derived exosomes on the development of bovine cloned embryos. PLoS One 12(3): e0174535
Ranjbaran A, Latifi Z, Nejabati HR, Abroon S, Mihanfar A, Sadigh AR, Fattahi A, Nouri M, Raffel N (2019) Exosome-based intercellular communication in female reproductive microenvironments. J Cell Physiol 234(11):19212–19222
Raposo G, Stoorvogel W (2013) Extracellular vesicles: exosomes, microvesicles, and friends. J Cell Biol 200(4):373
Record M, Subra C, Silvente-Poirot S, Poirot M (2011) Exosomes as intercellular signalosomes and pharmacological effectors. Biochem Pharmacol 81(10):1171–1182
Ronquist G, Brody I (1985) The prostasome: its secretion and function in man. Biochim Biophys Acta 822(2):203–218
Saadeldin IM, Kim SJ, Choi YB, Lee BC (2014) Improvement of cloned embryos development by co-culturing with parthenotes: a possible role of exosomes/microvesicles for embryos paracrine communication. Cell Reprogram 16(3):223–234
Saez F, Frenette G, Sullivan R (2003) Epididymosomes and prostasomes: their roles in posttesticular maturation of the sperm cells. J Androl 24(2):149–154
Segura-Benítez M, Carbajo-García MC, Corachán A, Faus A, Pellicer A, Ferrero H (2022) Proteomic analysis of extracellular vesicles secreted by primary human epithelial endometrial cells reveals key proteins related to embryo implantation. Reprod Biol Endocrinol 20:1–16
Segura-Benítez M, Bas-Rivas A, Juárez-Barber E, Carbajo-García MC, Faus A, De Los Santos MJ, Pellicer A, Ferrero H (2023) Human blastocysts uptake extracellular vesicles secreted by endometrial cells containing miRNAs related to implantation. Hum Reprod 38(8):1547–1559
Su Y, Li Q, Zhang Q, Li Z, Yao X, Guo Y, Xiao L, Wang X, Ni H (2022) Exosomes derived from placental trophoblast cells regulate endometrial epithelial receptivity in dairy cows during pregnancy. J Reprod Dev 68(1):21–29
Swann KM (2014) Effects of ovarian stimulation on oocyte development and embryo quality. University of Nottingham
Tan Q, Shi S, Liang J, Zhang X, Cao D, Wang Z (2020) MicroRNAs in small extracellular vesicles indicate successful embryo implantation during early pregnancy. Cells 9(3):645
Tan Q, Shi S, Liang J, Cao D, Wang S, Wang Z (2021) Endometrial cell-derived small extracellular vesicle miR-100-5p promotes functions of trophoblast during embryo implantation. Mol Ther Nucleic Acids 23:217–231
Taylor E, Gomel V (2008) The uterus and fertility. Fertil Steril 89(1):1–16

Théry C, Ostrowski M, Segura E (2009) Membrane vesicles as conveyors of immune responses. Nat Rev Immunol 9(8):581–593

Traver S, Assou S, Scalici E, Haouzi D, Al-Edani T, Belloc S, Hamamah S (2014) Cell-free nucleic acids as non-invasive biomarkers of gynecological cancers, ovarian, endometrial and obstetric disorders and fetal aneuploidy. Hum Reprod Update 20(6):905–923

Wang X, Tian F, Chen C, Feng Y, Sheng X, Guo Y, Ni H (2019) Exosome-derived uterine microRNAs isolated from cows with endometritis impede blastocyst development. Reprod Biol 19(2):204–209

Wang X, Li Q, Xie T, Yuan M, Sheng X, Qi X, Xing K, Liu F, Guo Y, Xiao L (2021) Exosomes from bovine endometrial epithelial cells ensure trophoblast cell development by miR-218 targeting secreted frizzled related protein 2. J Cell Physiol 236(6):4565–4579

Welsh JA, Goberdhan DC, O'Driscoll L, Buzas EI, Blenkiron C, Bussolati B, Cai H, Di Vizio D, Driedonks TA, Erdbrügger U (2024) Minimal information for studies of extracellular vesicles (MISEV2023): from basic to advanced approaches. J Extracell Vesicles 13(2):e12404

White M, Bissiere S, Alvarez YD, Plachta N (2016) Mouse embryo compaction. Curr Top Dev Biol 120:235–258

Xie Y, Liu G, Zang X, Hu Q, Zhou C, Li Y, Liu D, Hong L (2021) Differential expression pattern of goat uterine fluids extracellular vesicles miRNAs during peri-implantation. Cells 10(9):2308

Yuan S, Wang Z, Peng H, Ward SM, Hennig GW, Zheng H, Yan W (2021) Oviductal motile cilia are essential for oocyte pickup but dispensable for sperm and embryo transport. Proc Natl Acad Sci 118(22):e2102940118

Zemelko V, Kozhukharova I, Alekseenko L, Domnina A, Reshetnikova G, Puzanov M, Dmitrieva R, Grinchuk T, Nikolsky N, Anisimov S (2013) Neurogenic potential of human mesenchymal stem cells isolated from bone marrow, adipose tissue and endometrium: a comparative study. Cell Tissue Biol 7:235–244

Zhang Y, Yu M, Tian W (2016) Physiological and pathological impact of exosomes of adipose tissue. Cell Prolif 49(1):3–13

Zhou W, Dimitriadis E (2020) Secreted MicroRNA to predict embryo implantation outcome: from research to clinical diagnostic application. Front Cell Dev Biol 8:586510

MIX
Papier aus verantwortungsvollen Quellen
Paper from responsible sources
FSC® C105338

If you have any concerns about our products,
you can contact us on
ProductSafety@springernature.com

In case Publisher is established outside the EU,
the EU authorized representative is:
Springer Nature Customer Service Center GmbH
Europaplatz 3, 69115 Heidelberg, Germany

Printed by Libri Plureos GmbH
in Hamburg, Germany